PSYCHOLOGICAL VULNERABILITIES
(A Case Study on Jail Inmates)

PSYCHOLOGICAL VULNERABILITIES
(A Case Study on Jail Inmates)

By
Roshan Lal Zinta
Dharmender Kumar

Neha Publishers & Distributors
4832/24, Prahlad Lane, S-207 Ansari Raod, Daryaganj, New Delhi-110002

Published by
Neha Publishers & Distributors
4832/24, Prahlad Lane, S-207 Ansari Raod, Daryaganj,
New Delhi-110002
Tel.: 011-43570976, 23278261
Email-nehapubdistributors@gmail.com

First Edition : 2013

ISBN : 978-93-80318- 32-5

Price : 525/-

Laser Typesetting by : Vinod Bhardwaj

Printed at: Vikas Computers, Delhi

PRINTED IN INDIA

PREFACE

In the present scenario the world is entering in the era of scarcity by intertwining in a spiral of decline and degradation. The 21st century being an age of anxiety and stress has pushed an individual towards economic fluctuation, unemployment, hatred and resentment by divided them on socio-cultural basis. For maintaining better survival the human being sometime does not hesitate to commit even heinous crime. The involvement of person in bad habits and their approaches towards drug addiction, substances abuses and their interaction with liquor mafia and fraudsters have further heightened criminal graph. Therefore, the misdemeanors in developed countries in general and the developing countries in particular have turned out to be a smoldering issue in the contemporary scenarios. For example Russia has become the world leader incarceration rated (558 citizens per 100,000) followed with America(519 per 100,000) whereas India has (23 per 100,000) those seems to be heightened on the basis of poverty, political instability, and the transition to an uncertain economic future.

It seems that the human longing in the form of waves, aspirations, and hopes have propelled them towards crime. Coleman rightly observed that in our complex and rapidly changing world human being are confronted with unfinished business (poverty, prejudices and discrimination); new business (alarming rise of drugs abuse and violent crime, energy and ecological crisis accelerating social change) ; and future business (building better world for human being). The crime is regarded as an essentially a social problem. Severe historical penalties include the breaking wheel, boiling to death, flaying, disembowelment, crucifixion, impalement, crushing, stoning, execution by burning, sawing, scaphism, necklacing, statesemberment Scythians, the Greeks, the Romans, the Persians and the Carthaginians.Until the 2nd century AD, torture was used only on slaves .Thc imprisonment today include: Punishment, deterrence, and rehabilitation.

The Inmates are still subject to censorship, limited conjugal and visiting opportunities, inadequate treatment and training facilities and almost unlimited transfers. They cannot vote in some provinces and have limited privacy and access to procedural due process. Prisoners theoretically have the same basic human rights as anyone else in society, but these rights are only as accessible to the inmate as current penitentiary administrative policies and procedures will allow. Inmates are an insular minority vulnerable to oppression and discrimination, not only from the "outside" but from the strict rules and values of the inmate power structure and its own codes of silence and loyalty. The emergence of prison gangs comprised of groups of inmates for group protection and power dominance has become yet another reality for the incarcerated offender. In United States penitentiaries in particular, the proliferation of prison gangs has reached crisis proportions. Many of these gangs have evolved to become organized crime syndicates involved in gambling, extortion, drug trafficking, prostitution and contract murder; and, where there is organized crime, violence prevails. There are approximately 3.00 lacs people in jail suffering from various mental health problems in United States. The serious problems for long term prisoners include travel distance for loved ones, privacy during visitation, privacy in cells and crowding.

More appropriately, the jail inmates suffer from autonomy, liberty, security, heterosexual and relationship. Other stresses behind the bars include separation from family, overcrowding, sensory deprivation, exposure to a high-density of hard-core offenders, and a variety of uncertainties, fears, and frustrations. Loss of social status, uncertainty of outcome of the trial, fear of punishment, staying in an unusual place like police station or jail and the financial upsets, harasses them. If the trial period is prolonged for months or years, which is very common, then the under-trial's mental condition deteriorates. The hard life in prison further aggravates the situation. Psychological factors such as frustration, hostility, and feelings of helplessness might be the cause or the consequence of criminal behavior, and in some cases, both. Incarcerated women have more physical, social, and emotional problems. Psychological torture includes deliberate use of extreme stressors and situations such as mock execution, shunning, violation of deep-seated social or sexual norms and taboos, or extended solitary confinement. Studies on the effects of prison crowding, long-term imprisonment, short-term detention, solitary confinement, death row, and the health risks associated with imprisonment provide inconclusive evidence regarding "the pains of imprisonment. The deterioration model holds that long term incarceration causes the deterioration of an inmate's personality and mental, emotional and physical well-being. Prisonization model holds that the longer inmates are incarcerated, the more "criminalized" and distanced they become from the values and behaviors of society outside prison walls.

For Mahatma Gandhi prison was a place for inner reflection; for Jawaharlal Nehru it was a place for literary activity; and or Sri Aurobindo it was a place for self-realization. Prisons have been used to limit the movement of persons labeled as 'dangerous', 'psychotic' or 'disturbed', a labeling process which began in the community, in the bad schools and continued through each stage of the criminal justice system. Prison conditions are hard on mental health because of overcrowding, violence, lack of privacy and meaningful activities, isolation from families and friends, uncertainty of life after prison, and inadequate health services. The physical, emotional and psychological prison environment faced by all federal inmates is determined by the prevailing beliefs and attitudes held by the criminal justice system and the general public concerning the appropriateness of certain types of punishment.

Roshan Lal Zinta
Dharmender Kumar

CONTENTS

SOCIO-PSYCHOLOGICAL VULNERABILITIES AMONG JAIL INMATES

The misdemeanor in developed countries in general and the developing countries in particular has turned out to be a burning issue in the contemporary scenarios. Russia stands at the top, to become the world leader in its rate of incarceration, where 558 citizens per 100,000 populations recline in prison system. India stands at the 21st place in the rate of incarceration with 23 per 100,000 populations; and 1.3 million Americans are behind bars. The U.S. rate of incarceration was 519 per 100,000 during the world and was 22 % during 1989. The rise of poverty, political instability, and the transition to an uncertain economic future might have played a crucial role in the increased crime rates. It seems that the human longing in the form of waves, aspirations, and hopes (Anandmurti, 2011) to achieve their physical, mental and spiritual currents of existence as well as the philosophy of La Dolce Vita of passing gentle life might have thwarted and strained the person to commit crime. In the present scenario the world is entering in the era of scarcity by intertwining in a spiral of decline and degradation (Mehta, 2011). For maintaining their better survival and proving them as fittest and gratifying all the need, human being sometime do not hesitate to commit even heinous crime. Beside this the person with bad habits and drug addiction (Hafeez, 2011), substances abuses as well as their interaction with liquor mafia (Takkar, 2011) and fraudsters (Singh, 2011) has shoved them criminal activities. Sometime the barbarous and shameful act is committed for honour (Divya, 2011).

It seems that various social and psychological factors have played pivotal role behinds criminal behaviour of the person wherein approximately 90% of murderers and 82% of other violent criminal act are performed by the male (HT, 2011). Further, persons mental state and disorder is also considered behind number of offenses that in the present scenarios is around 7% (Sinha, 2011). The crime is a major activity across the world with implications for poverty and the allocation of public and private resources. Coleman (1969) rightly observed that in our complex and rapidly changing world human being are confronted with unfinished business (poverty, prejudices and discrimination); new business (alarming rise of drugs abuse and violent crime, energy and ecological crisis accelerating social change) ; and future business (building better world for human being). Many feel that human race is quite simply doomed. Crucial problem of war, violence, group discrimination, ecological violation and uncontrolled technological and social change has affected the life of the people. Since the 21st century being an age of anxiety and stress has pushed an individual towards economic fluctuation, unemployment, hatred and resentment by divided the people based on socio-cultural communities that definitely has led individual to commit crime.

The economics of crime focuses on the effect of incentives on criminal behavior; the way decisions interact in a market setting; and the use of a benefit-cost framework to assess alternative strategies to reduce crime. This essay shows that most empirical evidence supports the role of incentives in the criminal decision: legitimate labor market experiences, sanctions including incarceration, and the risk of apprehension all influence decisions to engage in crime. By putting crime into a market setting, economic analysis highlights the difficulty of reducing crime through incapacitation: when the elasticity of supply to crime is high, one criminal replaces another in the market; and thus the importance of deterring crime by altering behavior. Most analyses show that "crime pays" in the sense of offering higher wages than legitimate work, presumably in part to offset the risk of apprehension. But some important facts about crime — long term trend increases and decreases; the geographic concentration of crime; the preponderance of men and the young in crime — seem to go beyond basic economic analysis According to (Freeman, 2007). According to Hafeez, (2011), there are certain sections of the society who are very dominant who even do not bother the decision of the Supreme Court. In India such group belongs to Khaps Panchayat that has ruled out Supreme Court observation that perpetrators of honour killing should be awarded death penalty (Deswal, 2011).

The growing prison populations in the U.S. are largely due to drug-related crime and drug abuse. Yet, relatively few inmates receive treatment, existing interventions tend to be short-term or non-clinical, and better methods are needed to match drug-involved inmates to level of care. Using data from the 1997 Survey of Inmates in State Correctional Facilities, a nationally representative sample of 14,285 inmates from 275 state prisons, we present a framework for estimating their levels of treatment need. The inequality also prevails in Himachal Pradesh a sobriquet "land of Gods" and on the basis of religion where social discrimination with low caste people is practiced overtly and covertly in rural and urban areas (Bhat, 2011). When there is a threat to the unity, integrity and sovereignty of India then the Supreme Court gives stringent decision (According to Times News Network, 2011). The high class and income status people are associated with honors related crime (Divya (2011; Kamal, 2010 that in fact is very shameful. Farther new cases and law in come people are incarcerated and No medical facilities is provided inside the jail (Sharma, 2011). There are certain sections of the society who are very dominant ruled out the decision of Supreme Court (Deswal, 2011). Sometime some innocent also got in conspiracy and involves in the terrorism and do riots (Sarwar, 2011).

Severe historical penalties include the breaking wheel, boiling to death, flaying, disembowelment, crucifixion, impalement, crushing, stoning, execution by burning, dism Punishment in Greece is the story of the brazen bull (Bronze bull heated then burning) proposed to Phalaris in the mid 6th century BC. And five pains (cutting nose, hand, foot, castrated and waist) in ancient China. The crucification (tied and nailed in cross) in Scythians, the Greeks, the Romans, the Persians and the Carthaginians (Napoleon, B. 1961). Deliberately painful methods of execution for severe crimes were taken for granted as part of justice until the development of Humanism in 17th century philosophy, and "cruel and unusual punishment" came to be denounced in the English Bill of Rights of 1689. The Age of Enlightenment in the western world further developed the idea of universalhuman rights. The adoption of the Universal Declaration of Human Rights in 1948 marks the recognition at least nominally of a general ban of torture by all UN member statesemberment, sawing, scaphism, or necklacing (Edward, 1985).

Historically, imprisonment was based on punishing those who wronged society, by inflicting suffering of the body – similar to the pound of flesh depicted within Shakespeare's Merchant of Venice. In contrast to this concept, today's imprisonment is no longer simply intended as an acute form of corporal punishments, but a method by which to work on a person's mind as well as his body, through three distinct areas – which include: Punishment, Deterrence, and Rehabilitation. These three unique areas, when interlinked into a single process are intended to allow society to remove criminals from a position where they may continue their criminal behaviour, place them into an institution that satisfies the masses who desire some form of retribution, persuade other would be criminals that such activities are not beneficial, and in time sculpt them into productive and law abiding citizens through positive psychological conditioning who may later be re-integration into society. Crime can be connected to poverty and incentive (Freeman, 2007). People with more expected money commit crime. Behind all types of crime certain Mafias are working incessantly those are responsible for sending innocent people towards the jails (Takkar, 2011).

For Mahatma Gandhi prison was a place for inner reflection; for Jawaharlal Nehru it was a place for literary activity; for Sri Aurobindo it was a place for self-realization. For the ordinary prisoner languishing behind bars for crimes committed or allegedly committed, who had succumbed to the fury of emotional passions, life in prison could be a place for a guilt-ridden or revengeful existence losing their worth in their own eyes as much as that in their family's and society's eyes. According to John Boone (1975), former Director of Corrections, State of Massachusetts, "Prisons have been used to limit the movement of persons labeled as 'dangerous', 'psychotic' or 'disturbed', a labeling process which began in the community, in the bad schools and continued through each stage of the criminal justice system. The result has been the destruction of thousands of lives. Prison is a taboo place that even most hardcore criminals dread to go. The prisoner shows psychotic symptoms like delusion, hallucination, chaotic thinking, and disruption of consciousness, memory and perception of the environment (DSM, 2000).

Prison conditions are hard on mental health because of overcrowding, violence, lack of privacy and meaningful activities, isolation from families and friends, uncertainty of life after prison, and inadequate health services (WHO, 1999). Psychological factors such as frustration, hostility, and feelings of helplessness might be the cause or the consequence of criminal behaviour and in some cases, both.The concept of Justice during the time of Jesus Christ, Romans, Jews Egyptians and many others cultures during that time included torture as part of their justice system. Prison reform measures should lead to some reduction in inmate's feelings of hostility, helplessness, and other negative emotions. They should be constructive and restorative.

Portraying Criminal Behaviour

Individual's criminal behaviour has been attributed to an inadequate ability to effectively deal with the severe stress, deprivation and low self-esteem and with peer pressure and codes of behaviours of group such as gangs. These factors can be severely compounded by the ingestion of drugs and alcohol which offer the users relief from emotional discomfort by impairing or eliminating normal level of awareness and impulse control. Under such circumstances intense feelings such as fear, frustration, anger, and greed can rapidly results in antisocial behaviour and the resultant incarceration in correctional institutions that in turn bring additional stress with the possibilities of further exacerbating these conditions (Fogel, 1993; Perkins, 1998). The crime is regarded as an essentially a social problem, and retribution as the object of improvement is discarded. Detention as an objective of imprisonment is also very limited in scope. Reformation of the offender is being regarded as an ultimate aim of the prison sentence.

People are most likely to commit a criminal act between the age of fifteen and twenty five years. Imprisonment is a method of dealing with people who commit crimes by confining them to a fortified boundary with certain strict rules for all - that is, the prison. Crime, like any other action of the body, is a manifestation of thoughts in the mind.

National debates on race and incarceration have usually focused on men rather than women and on justice and public safety rather than health. Yet, high rates of incarceration affect the well-being of women of color directly, in that incarcerated women are removed from their communities, they are placed in close proximity to a population of women with high rates of infectious and chronic diseases, and opportunities to link them to needed services are missed. Incarceration also affects families by separating women from their children, often forcing children into foster care and leaving them vulnerable to psychological, educational, and social problems. Prison eliminates current income and reduces future earnings by diminishing women's prospects for post release employment. Moreover, even a short stay in jail can lead to homelessness.

An equally important and less addressed in the literature, criminal justice policies aimed at men can also harm women. The disproportionate incarceration rates among Black and

Latino men affect women by reducing the pool of male partners who can contribute to family income, reducing overall employment rates in low-income communities, and diminishing men's ability to be consistent and present fathers. For some women, the incarceration of an abusive or criminally involved partner can offer safety. However, the vast majority of men return to their homes from jail or prison, and thus the failure of correctional facilities to provide most incarcerated men with substance abuse, mental health, or domestic violence services forces many women to make an unpalatable choice. They can either separate from a male partner who returns from jail or prison, thus reducing financial and emotional support, or take back a man with drug, violence, or psychological problems that may jeopardize the family's health and safety.

Inmates World

The prison is a place where people are caught, prosecuted for and remanded criminal activities are remanded. The main function of the prison services is to provide rehabilitative and reformative services to offenders in order to make them well-adjusted and become responsible citizens upon release. The behaviour of inmates undergoes a rapid negative transformation as a result of the psychological breakdown during incarceration. Little wonder then that the prisoners come out of the prison dejected, stigmatized, traumatized and hopeless (Agali, 2004; Onagoruwa, 2001). The physical, emotional and psychological prison environment faced by all federal inmates is determined by the prevailing beliefs and attitudes held by the criminal justice system and the general public concerning the appropriateness of certain types of punishment.

Some maintain that in order for a punishment to be acceptable to the public, it must clearly demonstrate adverse effects. Others insist that the punishment of incarceration does not have to, and should not, be equated with harm and that the creation of a humane and effective prison environment requires the development of mechanisms with which to reduce deprivation of liberty (Wormith, 1985). Beliefs and attitudes about punishment can be complex and inconsistent. For example, a 1985 Gallup Poll showed that while over half of those polled felt prisoners had too many rights, 23% felt they had too few rights and almost two thirds felt that greater emphasis in prison should be placed on rehabilitation and community reintegration, even at the expense of security risk (Wormith, 1985). Many people therefore do not equate prisoners' rights with rehabilitation and reintegration. Unfortunately this inconsistency does not provide the policy makers with any clear guidance from the general public.

While it may be tempting to dismiss the harsh prison environment of the past as archaic and no longer in existence in the modern day, many of the attitudes and practices of over 200 years ago still exist.The inmates remain a disadvantaged group socially, legally, and politically. The sentence of panishment is no longer codified in the criminal law, but it is still, in effect, practiced when offenders are sentenced to terms in isolated institutions, often great distances from their families, friends and homes. This affects female long term prisoners greatly, as there are only five federal institutions for women. Many female federal inmates are not able to stay in their home provinces, and this causes stress on these women, particularly those who are mothers of young children. And although most forms of corporal punishment have been banned from Canadian institutions, solitary confinement, which is widely considered to violate basic human rights and dignity, is still extensively used as a form of discipline. One report produced for the Solicitor General of Canada pointed out that human isolation was found to have marked physical effects. For instance, the "sudden death" phenomenon has been recorded, whereby inmates die for no apparent pathogenic reason, but from a sense of hopelessness and despair resulting from feeling that they have no control over their circumstances or future (John Howard Society of Winnipeg, 1990).

The Inmates are still subject to censorship, limited conjugal and visiting opportunities, inadequate treatment and training facilities and almost unlimited transfers. They cannot vote in some provinces and have limited privacy and access to procedural due process. Prisoners theoretically have the same basic human rights as anyone else in society, but these rights are only as accessible to the inmate as current penitentiary administrative policies and procedures will allow. Inmates are an insular minority vulnerable to oppression and discrimination, not only from the "outside" but from the strict rules and values of the inmate power structure and its own codes of silence and loyalty. Whether male or female, young or old, inmates have to adjust to an environment in which relationships (sexual or platonic) with the opposite sex are denied them. In penitentiaries for men especially, homosexual sex is often viewed as a commodity which must be forced, bought or traded; new inmates have to find their place in the sexual order by developing a sexual identity and establishing a reputation in the prison "sex pressure game," which is designed to determine who is weak or strong (Hopper, 1990).

The emergence of prison gangs comprised of groups of inmates for group protection and power dominance has become yet another reality for the incarcerated offender. In United States penitentiaries in particular, the proliferation of prison gangs has reached crisis proportions. Many of these gangs have evolved to become organized crime syndicates involved in gambling, extortion, drug trafficking, prostitution and contract murder; and, where there is organized crime, violence prevails. For example, in Texas alone between 1979 and 1985, 62% of the 94 inmate murders recorded were committed by prison gangs. Overall, prison gangs also accounted for at least 50% of all problems and violence in United States penitentiaries (Fong & Buentello, 1990). There has been increased attention paid to the emotional, psychological and physical well-being of inmates in recent years, a sentence of imprisonment still means that an offender is relegated to a life isolated from the rest of society. According to Guy, Platt, and Bullock (2007) the mental health status of prisoners in an urban jail is not satisfactory. For example in Philadelphia , 2/3rd subjects were found psychiatrically disturbed and in a need of specific mental health treatment services and 34% identifiable as psychopathologic and 11% further need immediate inpatient psychiatric hospital care. Therefore it becomes very important to analyze the jail inmates socially and psychological those are associated behind their wellbeing.

Socio-Psychological Vulnerabilities among Jail Inmates

Crimes are acts that are forbidden and punished by law; these acts may threaten the well- being of the society, or injure any of its members. Many inmates manifest mental disorders in prison as a result of stresses of incarceration. The stresses behind the bars include separation from their family members, over crowding, sensory deprivation, exposure to a high-density of hard-core offenders and a variety of uncertainties, fear, and frustrations. The period of trial is of great stress to the individual. Loss of social status, uncertainty of outcome of the trial, fear of punishment, staying in an unusual place like police station or jail and the financial upsets harass the individual. If the trial period is prolonged for months or years which are very common, then the under trial's mental condition becomes bad. The hard life in the prison further aggravates the situation. The under trial's quality of life and subjective well-being are seriously affected by aforesaid conditions in the prison.

The environment is very oppressive and depressive. Being caged, reminds one of dogs or other pet animals that are kept chained. It is very different from the world outside, the world in which the prisoners are born and brought up. In one day, they are alienated from their family, friends, and the society. The shame of being 'caged', the worry regarding their family's welfare and their ability to cope with this change (sometimes, permanent for life term sentence), the dreadful loneliness and ghostly silence of being alone with minimal stimulation, depression, and boredom, all add to the stresses of incarceration. Life in prison leads to many problems. Most of the prisoners are ill equipped to cope with the intense stress of prison life. Research has revealed that apart from the criminal aspect, many inmates manifest mental disorders in prison as a result of the stress of incarceration. Krishna (1993) reports the presence of neuroticism, anxiety, extraversion, and morality guilt among adolescents who are high on delinquent behavior. Osofsky (1996) revealed the presence of certain psychological or personality factors exhibited by prisoners and the influence of these in creating greater stress within the prison. These factors are neuroticism, anxiety, aggression, hostility, and guilt. That effect the wellbeing of the prisoner.

Other stresses behind the bars include separation from family, overcrowding, sensory deprivation, exposure to a high-density of hard-core offenders, and a variety of uncertainties, fears, and frustrations. The period of trial is of great stress to the individual. Loss of social status, uncertainty of outcome of the trial, fear of punishment, staying in an unusual place like police station or jail and the financial upsets, harass the individual. If the trial period is prolonged for months or years, which is very common, then the under-trial's mental condition deteriorates. The hard life in prison further aggravates the situation. Psychological factors such as frustration, hostility, and feelings of helplessness might be the cause or the consequence of criminal behavior, and in some cases, both. There is all danger that in such an environment prisoners becomes worse than before.

In the UK over 60 percent of women prisoners have young children. Women are nearly always the primary care-givers and are often single mothers. At least 4,000 children are affected by their mothers' imprisonment. Only three percent of women have a child in prison with them. So children have to be farmed out. Most are living with their mothers prior to the woman's imprisonment and for 85 percent it is the first time they have been separated for more than a day or two. Siblings are often separated. There is an evidence that most children become withdrawn and depressed. One in four has difficulty sleeping or becomes physically ill. Seventy percent see their mother only once a month or less. Half of all babies under one year who are in care because their mothers are in prison are moved between 2 and 4 different homes. We know that it is vital for babies in the first year of life to form a strong attachment to one person. From this grows a sense of security and trust, self-esteem and the capacity to love. The enforced separation of mothers and babies is another form of violence against women and is an abuse of our power over children *(Caddle & Crisp, 1997).*

Incarcerated women have also numerous physical, social, and emotional health care needs, including specific needs related to their expressions of sexuality while in prison. This report describes the results of a participatory action research study with incarcerated women utilizing critical hermeneutic data analysis techniques. The public's view of sexuality between incarcerated women borders on the prurient and profane, this study suggests that women in prison continue to be sexual beings that come to "participate" in love and sex with one another based on their need for relationship and friendship. It is suggested that prison bureaucracies define women through a sexual lens, dually grounding their identities in the crimes they were sent to prison for and the perceived crimes of their sexuality. The need for nursing involvement and intervention with this marginalized and stigmatized population is discussed (Maeve & Katherine, 1999).

There are approximately 3.00 lacs people in jail suffering from various mental health problems in United States (Abramsky & Fellner, 2003). The proportion of prisoners with mental health problem is increasing. Sheer number of mentally ill inmates has transformed prison into facilities for the mentally ill. Mentally ill prisoners however do not have the same capacity to comply with prison rule as do other prisoners. Inmates on mental health roaster have higher infraction rates than other inmates. Offender with serious mental illness is 18.7% in jails (Lovell & Jemelka, 1996). But the mentally ill people are treated with punitive measures without knowing the cause of behaviour and efficacy of such treatment. Mentally ill persons are routinely punished under prison disciplinary system for rule inflation arising from their illness without regard to their actual culpability. The number of incarcerated people with severe mental illness has increased tremendously in few decades that prison may now be the largest mental health providers across the countries. Impact of this problem is worse on the life of the prisoner. Their thinking and emotional responses are impaired by the schizophrenia, bipolar diseases and depression. Mentally ill prisoner face inadequate services. There is an inherent tension between security mission of prison and mental health consideration. The formal and informal rule and regulation, code and conduct in prison reflect staff concern about security, safety, power and control.

Psychological torture also includes deliberate use of extreme stressors and situations such as mock execution, shunning, violation of deep-seated social or sexual norms and taboos, or extended solitary confinement. Because psychological torture needs no physical violence to be effective, it is possible to induce severe psychological pain, suffering, and trauma with no externally visible effects. In medical torture, medical practitioners use torture to judge what victims can endure, to apply treatments that enhance torture, or act as torturers in their own right. Josef Mengele and Shiro Ishii were infamous during and after World War II for their involvement in medical torture and murder. Pharmacological torture is the use of drugs to produce psychological or physical pain or discomfort. Tickle is an unusual form of torture which nevertheless has been documented, and can be both physically and psychologically painful. The ancient Greeks and Romans used torture for interrogation. Until the 2nd century AD, torture was used only on slaves (with a few exceptions).

The well-being, therefore, is a complex construct that is concerned with optimal experience and functioning. Current research on well-being has been derived from two general perspectives: the hedonic approach, which focuses on happiness and defines well-being in terms of pleasure attainment and pain avoidance; and the eudaimonic approach, focuses on meaning and self-realization and defines well-being in terms of the degree to which a person is fully functioning. These two views have given rise to different research foci and a body of knowledge that is in some areas divergent and in others as complementary. New methodological developments concerning multilevel modeling and construct comparisons are also allowing researchers to formulate new questions for the field. This review of research from both perspectives concerning the nature of well-being, its antecedents, and its stability across time and culture (Ryan & Deci, 2001).

Hence, the general well being may be defined as the subjective feeling of contentment, happiness, satisfaction with life experiences and one's role in the world of work. Thus defined and conceptualized the general well being may show some degree of positive correlation with quality of sense of achievement and negatively with neuroticism, psychoticism and other such related variables However the degree of overlap with such variables should not be considered high if this concept as separate entity is to be considered as a valid type. Also it should show relative stability over time. So it is very difficult to provide a definite list of traits. More appropriately, the well-being comprises sense of achievement utility, belonging ness, and non-distress, dissatisfaction or worry etc. those in some extent are very difficult to evaluate objectively. The w***ellbeing as a noun refers to the*** state of being well, happy, and prosperous and welfare. Finally, the global well-being involves present life satisfaction, global self-esteem and self-efficacy.

The term 'psychological well-being' as used in this study refers to a prisoner's perception of the overall quality of his/her life. It is synonymously used with such terms as "overall life satisfaction", inner sense of security" and "psychological adjustment". These perceptions are essentially affective evaluations which the prisoners attach to the external or the internal aspects of their lives. Such evaluations, according to Odu (2000) determine, to a large extent, the trend of an individual's life. As prisons are a high stress environment facilitating depression, aggression, and sexual offences; it is an ideal place to introduce spirituality as a coping strategy. Prison schedules are structured in their setting and functioning, with limited time pressures. The incarcerated have lot of free time which most of them find the hardest to cope with.

Objectives of the Present Study

The objectives of the present study is to appraise psychological vulnerabilities among the jail inmates and non-inmates of Shimla. Further the study intends to assess and compare the psychological wellbeing among married and unmarried people in general as well as married jail non-inmates and inmates in particular. The intention behind their exploration is that the married people seems to be more attached to their families, wife and the children compared to the unmarried counterpart. Beside this the study has also tried to explore the qualitative information about the jail inmates who have incarcerated for the short period of time, moderate period of time and long period of time. Further the study also intends to see the relationship between birth order among the brother and its relation with criminality and wellbeing. But the analysis performed in the present study dealt only with married and unmarried jail inmates and non-inmates. The detailed objectives are as follows:-

1. To identify the difference between married and unmarried jail inmates and non-inmates on the depression and happiness parameter.
2. To identifies the difference between married and unmarried jail inmates and non-inmates on the parameter of anger.
3. To identify the difference between married and unmarried jail inmates and non-inmates on satisfaction with life parameter.
4. To identifies the difference between married and unmarried jail inmates and non-inmates on general health measure.
5. To identify the difference between married and unmarried jail inmates and non-inmates in religiosity measure.
6. To identify the difference between jail inmates and non-inmates in alcohol, smoking, and substance involvement measure.

IMPACT OF LONG-TERM INCARCERATION ON WELLBEING

Impact of Long-Term Incarceration on Wellbeing

The imprisonment necessitates a substantial curtailment of an individual's freedom and many other basic rights. The deprivation is an inherent feature of being incarcerated. In a study of long term inmates in Missouri, Sabbath and Cowles (1992) found that the most serious problems for long term prisoners include travel distance for loved ones, privacy during visitation, privacy in cells and crowding. These problems are indicative of various forms of deprivation. An earlier study, carried out by Timothy Flanagan to examine the attitudes and perspectives of long term inmates (more than 10 years) listed five problems from most to least were found as missing social life, worrying about how they will cope when released, feeling that their lives are being wasted and feeling sexually frustrated They also reported loss of relationships with family and friends outside the prison was consistently mentioned as the most serious deprivation. Some long term inmates cut themselves off from these relationships as a means of avoiding the anxiety and despair that accompany separation. However, for the majority of inmates who do not use this strategy, family ties become a two-edged sword over the years, providing encouragement and support and at the same time, making it more difficult to serve time (Flanagan, 1980).

Most of the inmates reported that imprisonment had not seriously threatened their emotional well-being. Fellow inmates were found with the problems of their own, family members are not seen as alternatives and institutional staffs are viewed as unconcerned (Flanagan, 1980). Another examination of the "pains of imprisonment," carried out earlier than the Flanagan study, was that of Graham Sykes. Four basic deprivations presented in his work concerned liberty, autonomy, personal security and heterosexual relations (Sykes, 1966). Sociological research of recent years has shed light on the effects of these deprivations.This research examined the structural and social psychological determinants of prisonization among a sample of 239 male inmates in a maximum security prison. The study examined the process of prisonization more fully than has been done in the past by developing a theoretical model that integrates measures of importation, deprivation, and inmate self-conceptions. To date, the self-concepts inmates bring with them and sustain in prison have been a neglected aspect in the study of prisonization. The results of this study demonstrated that while deprivation model variables were shown to be the better predictors of prisonization, certain importation model variables were not without significant impact and the two models do explain more variance together than either one separately. Measures of self-conception were of limited value in predicting prisonization (**Brent & David, 1999**).

Deprivation of Autonomy

The world of the inmate is characterized by a multitude of rules and commands designed to control his or her behaviour. Yet, some argue that the inmate is not much worse off than the individual in the free community who is regulated in a great many aspects of his or her life by the dictates of custom. However, regulation by a bureaucratic staff is felt far differently than regulation by custom. Most prisoners express an intense hostility against their far-reaching dependence on the decisions of corrections officials, which is what makes their restricted ability to make choices one of the major deprivations of imprisonment (Sykes, 1966). Long term inmates often lose their sense of self efficacy once autonomy is taken away. Offenders are told where to live and when and what to eat, they are required to wear regulation clothing, perform certain jobs and follow numerous rules (Santos, 1995). Self motivation and personal achievement are neither facilitated nor reinforced among inmates.

The Rigid sometimes incomprehensible rules have always been basic features of incarceration. Inflexibility and unresponsiveness to the concerns of prisoners often results from bureaucratic indifference, whereby events which seem important or vital to those at the bottom of the heap are viewed with an increasing lack of concern with each step upward. The rules, commands and decisions that are imposed on inmates are not accompanied by explanations, as many corrections officers feel that they do not need to justify their demands and actions; inmates are to do what they are told and not ask questions. Thwarting the inmate's ability to make choices and refusing to provide an explanation for prison rules and regulations involves a profound threat to the inmate's self image by reducing the inmate to the weak, helpless, dependent status of childhood (Sykes, 1966).

Loss of autonomy can also entail a serious threat to the inmate's self image as a fully accredited member of adult society. Public humiliation, enforced respect and deference, the finality of authoritarian decisions, and the demands for certain conduct because it is in the individual's best interest are all features of childhood helplessness in the face of a superior adult world. This may be irksome and disturbing to a child, but for the adult who has escaped such helplessness with the passage of years, being thrust back into such helplessness could prove even more painful (Sykes, 1966). Treating inmates as if they were children is contrary to the best interest of society: when long term prisoners are released they may have lost the ability to make decisions for them and are less likely to be able to live productive lives in the community.

Deprivation of Liberty

Few of the conditions imposed on inmates are as severe the loss of liberty. Inmates, particularly those serving their sentences in maximum security facilities or in special handling units, must live in a world where their freedom of movement is rigidly restricted and regulated. For long term inmates, liberty may be deprived for substantial portions of their lives and can have a serious effect on their mental health. The inmate's loss of liberty occurs at two levels; first by confinement to the correctional institution and second, by confinement within the institution. At the first level, offenders are cut off from family, relatives and friends, producing what can be a painful deprivation and frustration in terms of lost emotional relationships, loneliness and boredom. Most of their waking hours are spent within the confines of their cells.

The second level of loss of liberty occurs when offenders are put in solitary confinement. Sensory stimulation is quite limited and inmates may, in some facilities, be allowed to read a book while in their cells or exercise for one half hour per day outside of the cell. Studies of the effects of solitary confinement generally show that short periods in isolation do not have detrimental repercussions on the mental health of inmates. Prolonged periods of isolation may have negative impact on inmates as indicated by studies conducted by Cormier and Williams (1966) and Grassian (1983)(cited in Bonta & Gendreau, 1990), but because adequate controls were not included in these studies, more studies need to be conducted before a definitive conclusion on the effects of solitary confinement can be made. Many mental health experts would argue that solitary confinement is, for the majority of offenders who spend long periods in solitary, a psychologically damaging punishment. There are varieties of psychological symptoms ranging from "memory loss to severe anxiety to hallucinations to delusions and, under the most severe cases of sensory deprivation, people go crazy" (CNN, 1998). The almost complete loss of liberty that solitary confinement entails is dehumanizing and may hurt the inmate's potential for rehabilitation.

Deprivation of Security

The prison inmates are deprived from security. The offender is placed into prolonged proximity with other inmates who in many cases have a long history of violent and aggressive behaviour. It is a situation which has proven to be anxiety provoking for even the hardest of recidivists. Regardless of the mutual aid and support which may flourish in the inmate population, there are a sufficient number of offenders within this group of offenders to deprive the average inmate of the sense of security which comes from living among people who can be reasonably expected to abide by the rules of society (Sykes, 1966). The loss of security arouses acute anxiety, not just because violent acts of aggression and exploitation can take place, but also because such behaviour constantly calls into question the individual's ability to cope in prison and hinder their abilities to live normally in the outside world. The thoughts of a long term inmate beginning a 45 year sentence in an American prison illustrate these problems:

Another threat to the personal security of inmates is the possibility of contacting the HIV virus from another inmate through sexual activity, sharing syringes used to inject intravenous drugs and through unsanitary tattooing practices. Correctional Service Canada (CSC) statistics indicate that in 1997, the known cases of HIV in federal institutions totalled 158 and the number of unknown cases were thought to be much higher (CSC, 1998). Contact with HIV infected prisoners entails a risk when bodily fluids are shared, and because most inmates do not know the HIV status of fellow inmates, they may engage in high-risk activities with little fear of HIV infection. Others might become fearful of infection, particularly those who have been sexually assaulted.

Deprivation of Heterosexual Relationships

Some researchers have suggested that male inmates undergo a reduction of the sexual drive and that the sexual frustrations of inmates are, therefore, less significant than might be expected. However, these reports were largely based on the accounts of men imprisoned in concentration camps or similar extreme situations where starvation, torture and physical exhaustion reduced life to a struggle for survival or left the captive deep in apathy (Sykes, 1966). In addition to the physiological effects of sexual frustration, Sykes cited possible psychological problems created by the lack of heterosexual relationships for male inmates. A society composed exclusively of men tends to generate anxieties in its members concerning their masculinity, regardless of whether or not they are coerced, bribed or seduced into an overt homosexual liaison. Latent homosexual tendencies may be activated in the individual without being translated into open behaviour and yet still arouse strong guilt feelings at either the conscious or unconscious level.

The deprivation of heterosexual relationships carries with it another threat to the prisoner's image of himself-more diffuse, perhaps, and more difficult to state precisely and yet no less disturbing. The inmate is shut off from the world of women which by its very polarity gives the male world much of its meaning. Like most men, the inmate must search for his identity not simply within himself but also in the picture of himself which he finds reflected in the eyes of others (Sykes, 1966, p. 72). Paradoxically, many inmates who consider themselves to be heterosexual assert their masculinity not by suffering through the frustrations of abstinence, but by engaging in homosexual activities. In some institutions for male offenders, sub-cultural norms exist allowing "real men" to have homosexual relations without having their heterosexual identity challenged, as long as they always take the penetrative role. Receptive males, many of whom are unwilling participants in the sexual activity, are stigmatized and may be subject to prostitution and rape within the institution (Donaldson, 1990). The combination of sexual frustration and the need to maintain one's masculine image while facing a long period of incarceration with only members of the same sex leads many male inmates to acquire sexual gratification from other men through persuasion, bribery, coercion or force.

The homosexual relationships among female long term inmates emphasize the satisfaction of emotional needs in addition to the desire for sexual release. In a study of females sentenced to life in a British institution, Genders and Player found that "considerable consternation was expressed by the women about the lack of "normal contact" with men and the implication of such deprivation for their future sexual relationships" (1990, p. 124). Most female inmates do not consider themselves lesbian, but turn to other women because men are unavailable to fulfill their emotional and physical needs. This can lead to confusion and anxiety when they return to the outside world where men are available to them. The women also turn out to be lesbian (Genders & Player, 1990, p. 124). Deprivation of heterosexual relationships has a profound impact on both male and female inmates. Homosexual relations are the only means for participatory sex in prison, and many; female inmates in particular, undergo changes in identity and self perception. Some male prisoners are spared this change and are allowed to maintain their heterosexual "real man" status, but oftentimes this requires the victimization and abuse of other prisoners.

Theories and Models of Imprisonment

Until the 1980s, most studies and theories concerning the long term effects of incarceration were aimed at substantiating the belief that lengthy incarceration inevitably lead to a systematic physical, emotional and mental deterioration of the long term inmate. However, more recent studies have suggested that not only is this deterministic premise too simplistic, but that the methodology applied in many of the studies provided little in the way of substantiated, empirical facts. According to Bonta and Gendreau, Studies on the effects of prison crowding, long-term imprisonment, short-term detention, solitary confinement, death row, and the health risks associated with imprisonment provide inconclusive evidence regarding "the pains of imprisonment." (1990, p. 75)

Cornerstone theories about the effects of imprisonment as deterioration, deprivation and prisonization have been challenged and found lacking in reliable evidence based on systematic observations of behaviour. Despite the inconclusiveness of the research findings, an exploration of the literature on the effects of long term incarceration gives us a great deal of insight into the prison experience. Understanding the impact of prolonged imprisonment is essential, as the vast majority of long term inmates will one day be released. It is therefore imperative that we address the potentially damaging aspects of incarceration in order to improve each offender's chance for rehabilitation, as the protection our communities hinges on treatment and reintegration of offenders.

Deterioration Model

The deterioration model holds that long term incarceration causes the deterioration of an inmate's personality and mental, emotional and physical well-being. Clinical and psychiatric case studies have long suggested that imprisonment can be devastating, at least for some offenders. For example, some studies into functional "psycho-syndrome" have shown inmates with defects in cognitive functions, such as loss of memory and a general clouding of comprehension and ability to think; other defects included emotional problems (apathy and rigidity), problems in relating to others (infantile regression and increased introversion), and the appearance of psychotic characteristics (obsessions, loss of reality contact) (Zamble & Porporino, 1988). However, other studies which applied even more stringent methodology provided no consistent findings of significantly quantifiable psychological deterioration. There have been numerous attempts to use traditional measures to assess the effects of imprisonment on personality, but no clear conclusions were reached from these studies. Studies using the self esteem dimension of personality as a measure resulted in contradictory findings; some studies showed self esteem increased after some period of incarceration, while other studies indicated no change in self esteem (Zamble & Porporino, 1988).

Several studies, such as those carried out by a team of psychologists at Durham University in England, attempted to assess the effects of imprisonment using comprehensive batteries of psychological measures with groups of inmates who varied in the amount of time they had served. It was concluded that there was no over-all deterioration in perceptual-motor or cognitive functioning in connection with duration of imprisonment. Furthermore, there were no consistent changes in attitudes or personality resulting from length of imprisonment. Other similar studies in Germany and Canada showed that while a large proportion of inmates showed signs of depression and emotional withdrawal, very few differences were found based on length of incarceration. Furthermore, bitterness and expressions of demoralization by the prison environment, such as sleep disturbance and loss of appetite, were most evident among inmates who had served the least time (Zamble & Porporino, 1988). Nonetheless, the evidence from these studies still indicates that the gross psychological deterioration suggested by the traditional deterioration model is not an inevitable consequence of long term incarceration and that imprisonment is not generally or uniformly damaging (Zamble & Porporino, 1988). Clearly, more research needs to be done into the effects of incarceration on the individual level, as each offender will respond to their environment differently. The long term inmates sentenced for different lengths of time in different institutions are subject to varying degrees of deprivation and isolation, and programming available for these offenders will also vary. Thus, future studies undertaken to assess deterioration in prisons need to be strictly controlled for these factors.

Prior to the rise of HIV and other life threatening infectious diseases like hepatitis C, physical deterioration in prisons was not definitively established in the social-scientific literature. Bonta and Gendreau (1990) argue that this is in part due to the widespread availability of medical services within correctional institutions. In general, prison food is adequate and inmates often eat a healthier variety of foods than they would on the "outside." In the last decade, however, it has become apparent that inmates are at a much higher risk than the general public for contacting HIV and hepatitis B and C. From January to July of 1995, 18 new cases of active hepatitis B and 200 new cases of hepatitis C arose in federal correctional facilities (Malkin, 1995). In that year, after studies were conducted at three penitentiaries, it was estimated that between 28% and 40% of federal inmates were hepatitis C positive. In the month of August 1995, 152 federal inmates had been classified as HIV positive, up 40% since April 1994 (Jürgens, 1996). The number of HIV cases in federal prisons has continued to rise. In 1997, there were 158 known cases of HIV (CSC, 1998). It has been reported that prison infection rates for HIV is 10 times that of the general Canadian population (Jürgens, 1996; CSC, 1998). The longer a person remains in prison thus, the more likely they are to become infected with the HIV virus.

2.2.2 Prisonization Model

The prisonization model, first developed in 1940, holds that the longer inmates are incarcerated, the more "criminalized" and distanced they become from the values and behaviours of society outside prison walls. A process involving changes within the individual inmate, prisonization results in the inmate increasingly acquiring the values, standards and behaviour patterns of the other inmates; imprisonment causes prisonization, which in turn results in the inmate assuming criminal role identities (Zingraff, 1975). Two variations on the prisonization theory have emerged. Some sociologists have argued that it is pre-imprisonment attitudes and behaviour patterns and the duration of involvement with criminal value systems prior to incarceration which are the crucial determinants of prisonization (Irwin, 1970, Irwin & Cressey, 1962, Thomas & Petersen, 1977, Zamble & Porporino, 1988). Others have argued it is primarily factors within the corrections institution which determine the prisonization process. They suggested that the degree of prisonization could be affected by such factors as length of time incarcerated, interpersonal ties with other criminals, proportion of time served, social role adaptation of the inmate, post-release expectations of the inmate, degree of alienation from society, degree of alienation from the institution and self-concept of the inmate (Zingraff, 1975).

Considerable research has been carried out on both varieties of the prisonization theory. Generally, studies into the theoretical links between prisonization and the various pre- and post-imprisonment factors have found weak and inconsistent relationships (Bowker, 1977, Hawkins, 1976, Thomas, 1977, Zingraff, 1980, cited in Zamble & Porporino, 1988). Other findings have also casted doubts on the prisonization theory. Although prisonization has been consistently related, at least theoretically, with decreased likelihood of post-release success, several studies have indicated the opposite. Inmates who subscribed to the inmate code and adjusted poorly to the prison structure were found less likely to be recidivists than the less prisonized inmates (Kassebaum, Ward & Wilner, 1971, Zamble & Porporino, 1988). Other studies showed that rebellious inmates were less handicapped during the initial stages of transition into the community (Goodstein, 1979, Zamble & Porporino, 1988).

More recently, studies are increasingly indicating that the prisonization model has not clarified why and how inmates adapt in particular ways during confinement. The concept of prisonization focuses on explaining uniformity of behaviour rather than the individual variation. Incarceration affects individuals in different ways and while some inmates may exhibit similar "prisonized" behaviour, they may still adapt quite differently in other ways. Recent studies into prisonization have suggested that in order to understand varying reactions to imprisonment, a much finer analysis is needed than has been provided by the prisonization theory. Some researchers maintain it may be more useful to view prisonization as an attitudinal factor which combines with other variables to affect adaptation to incarceration rather than as a primary method of inmate adaptation that could be predicted (Zamble & Porporino, 1988).

Attachment Theory

This theory was given by John Bowly in (1969). His theory is based on the bond established between mother and infant. According to him care giving experiences are internalized as working models that not only serve as a prototype for future relationship with significant others but also provide unwritten rules for how one experience, expresses and copes with distressing emotions. This theory is based on three tenets: proximity to other, proximity maintenance depends on the responsively of other person to ones attachment need and third is experience with significant other. Childs internalization of relationship with an attachment figure create a secure base of support and comfort in time of distress and facilitates the child's exploration of widest world (Ainsworth, 1989; Bowlby, 1969, 1973, 1980, 1982). Recent theories have suggested this relational process as the circle of security (Marvin, Cooper, Hoffman & Powell, 2002) in which a Childs move from secure base to exploration and then back to safe heaven following a rupture. Contemporary theory and research recognize the interconnection between attachment, care giving, and exploration for adults (Freeney & Collins, 2004; George & Solomon, 1999). Similar to child's adult attachment system is activated in times stress or novelty and adult turn to their attachment figure to alleviate this distress and regulate their affect (Hazan, Gur-Yaish & Campa, 2004). Bowlby and Ainsworth further proposed that young children internalize early attachment processes as internal working models of attachment which serves as relational templates with continuity throughout the life span (Hesse, 1999; Fonagy, 2001). Ultimately adult navigates their relational world influenced by either secure or insecure internal working model developed in their early work. Secure adult attachment relations are characterized by the capacity to relate to others in mutual reciprocal manners to provide coherent narratives about relationships, and to sustain continuity of connections (Basham & Miehls, 2004; Hesse, 1999).

Further Fisher and Crandle (1977) suggest that couples who are securely attached "shift freely between the dependent and depended upon position" with an open expression of need for comfort and contact as well as an open reception of that contact. In other words the attachment system of secure adult partners can be appropriately activated and de-activated in a flexible manner depends on the emotional need of an individual at any given time. In contrast insecure individual may be dismissing, pre-occupied or unresolved/ disorganized in their state of mind about attachment (Hesse, 1999). It leads them towards bi-directionality, flexibility or mutuality common among secure partner. Recent research suggest that adult couples may create way of relating that are specific to the relationship and perhaps different from individuals global model of attachment or state of mind about early childhood attachment relationship.

Social Theory

The main objective of social theory (Banton, Clifford, Erosh, Lousada & Rosenthal, 1985, Jones & Moon, 1987) is to understand how social or society enters into the life of an individual and construct him socially. This theory identifies three linked practices important in this process. The first is economic practice, second is political practice and the third is ideological practice. Moreover, the social theory reveals that people are subject to economic, political and ideological forces, which are more powerful than other conscious factors affecting individuals. It also talks about the process of Subjectification refer to a dual function of "socialist therapy". First, it aims to reveal the individual, the way she/he is subject to unconscious forces and the second aim is to pursue the analysis of these forces until their social determinants are uncovered. In this way distinguishing characteristic of social therapeutic practice is to work out the structure of power relations in the individual power relations to Self-control. Furthermore, those power relationships would determine the need for change and how that change can be brought out.

This theory also distinguishes 'remembered' and 'real' community. It is through remembered community that the contradictions and antagonisms develop inside the individual. The 'remembered community' reflects a paradigm of community life that is reinforced or strengthened through media, politicians and welfare workers. The structure of the state also influences community in various ways. On this basis the distinction or classifications are made between rich and poor, different classes or castes. The control of the state extends to an extent that there is virtually management of subjectivity through mental health industry. This theory calls for development of real community alternative, which involves various processes of socialization.

Social Cognitive Theory

This theory was formulated by Albert Bandura (1961, 1977, 1986, 2001, and 2004). Basic purpose of social cognitive theory was to understand and predict individual and group behaviour; to identify methods in which behaviour can be modified or changed and to use intervention aimed at personality development, behaviour pathology and health promotion. The theory defines human behaviour as triadic, dynamic, and reciprocal interaction of personal, behaviour and the environmental factors. It contends that behaviour is largely regulated antecedently through the cognitive processes.

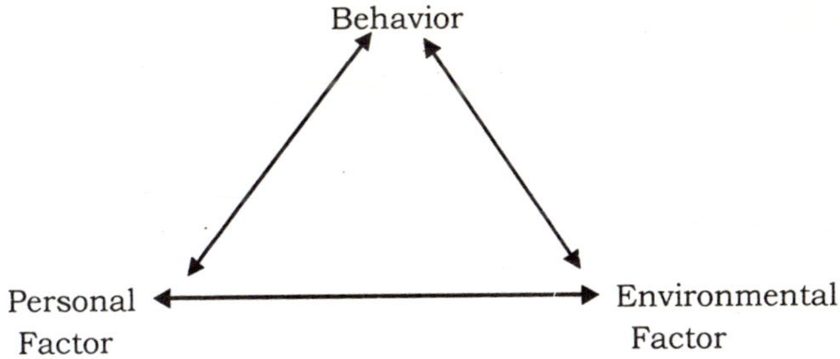

Figure 1: Conception of triadic reciprocal determinism

(Source: Albert Bandura, 2001)

In reciprocal determinism there is interaction between environment, personal factors and behaviour. The symbolizing capability dealt with most external influences that affect behaviour through cognitive processes and the symbols (e.g. images or mental pictures or words) serve as the mechanism for thought. Vicarious processes refer to human ability to learn not only from direct experience but also from the observation of others. Forethought is a person's capability to motivate them and guide their action anticipatorily (Bandura, 1989). Self-regulatory systems are an internal control mechanism that governs what behaviour is performed and the self-imposed consequences for that behaviour. The self-reflective capability enables people to analyze their experiences, think about their own thought processes, and alter their thinking accordingly (Bandura, 1986, 2001, 2004).

The response consequences of behaviour are used to form expectations of behavioural outcomes. It gives strong emphasis on one's cognition by suggesting that the mind is an active force that constructs one's reality, selectively encodes information, performs behaviour on the basis of value and expectations and imposes structure on its own actions. Key concept in this theory is reciprocal determinism, symbolizing capability, vicarious capability, forethought capability, self-regulatory capability, and self-reflective capability. Social cognitive theory involves four types of agencies named as personal where cognitive, motivational and affective and choice processes through which it is exercised to produce given effect. The manner in which human agency operates has been conceptualized in at least three different ways-as autonomous agency, mechanical agency or emergent interactive agency. In the concept of agency self-referent processes are epiphenomenalism by products of conditioned responses that do not enter into the determination of action. Successful functioning requires and agentic blend of these different modes of agency (Pajares, 2002).

The Coping Model

Generally speaking, it makes little sense to search for the psychological effects of incarceration without acknowledging that these effects may vary widely among individuals. This fac\t is recognized by proponents of the coping theory who examine individual differences in how inmates adapt to their environment. Their research is unlike most studies into the effects of incarceration, which are focussed primarily on finding a general, uniform set of psychological effects upon which to base predictions about long term inmate behaviour. Operating on the premise that how individuals cope with problems is more important than the frequency and severity of the problems experienced, the coping theory focusses on the interaction of the personal and environmental factors involved as inmates adapt to a life of incarceration.

An example of how coping behaviours would represent an interaction between the inmate and the penitentiary environment could be the comparison of two individuals, both of whom are facing a long sentence in the same institution. Both would experience the same environment characterized by restrictions and deprivations and both would likely be confronted with events during incarceration which are beyond their control. However, as a result of individual history, attributes, beliefs and coping capabilities, one person could interpret the lack of control as the result of personal inadequacy, while the other could interpret it as continuing abuse by others. While the first may sink into depression, apathy and withdrawal, the second might become resentful, angry and rebellious in an attempt to counter the control.

The way the two individuals deal with their long sentences could also determine how they are each affected by the environment. While one might cope with the stress of long confinement by avoiding all thoughts of the future, the other may cope by finding a safe and comfortable behavioural niche within the institution. The first could take on the behaviour and values of the other inmates and be seen by outsiders as acting impulsively and carelessly; the second might have much weaker ties to the inmate subculture. The behaviour each exhibits would in turn affect the way each is seen by staff and other inmates, and their subsequent treatment would differ. This would further affect the emotional responses and the ways the two individuals appraise their environments, with subsequent behaviour then being affected by each factor continuously (Zamble & Porporino, 1988).

A Canadian study was carried out to test the coping hypothesis by observing, over a period of 16 months, the coping and adapting behaviour of inmates serving sentences similar in length in the same penitentiary. Conditions for the incarcerated are very different from those on the outside. By its nature, imprisonment requires that inmates live apart from their families, live together in groups that would not otherwise exist and be seriously constrained in their choices. As a result the inmate's world is socially and psychologically artificial. A primary objective of this study was to further understand what being incarcerated does to the inmate and to be able to make some conclusions about the general effects of imprisonment.

Promoting Wellbeing among Jail Inmates is an ardent talk

The religion has long been considered by some to be a deterrent to crime because it promotes social control, encourages the development of moral character and the acceptance of societal norms and values (Davis, 1948; Erikson, 1966; Fitzpatrick, 1967). However, this notion was disputed by Hirshi and Stark (1969), who reported that church attendance and belief in supernatural sanctions were unrelated to self-reported delinquency; that religiosity and delinquency were virtually unrelated. More recently, analyzing the results of a study by Evans et al. (1995), it was summarized that involvement in religious activities may significantly decrease the risk of adult criminality, as it may for adolescent delinquency (Koenig et al, 2001, p. 183-184). "there is growing evidence that religion can be used as a tool to help prevent high-risk urban youths particularly from engaging in delinquent behavior. Religious involvement may help adolescents learn "pro-social behavior" that emphasizes concern for others' welfare. Such prosocial skills may give adolescents a greater sense of empathy toward others, which makes them less likely to commit acts that harm others. Similarly, once individuals become involved in deviant behavior, it is possible that participation in specific kinds of religious activities can help steer them back to a course of less deviant behavior and, more important, away from lives as career criminals". However, this correlation does not answer questions raised by increased inter-religious crimes that are the order of the day. The crucial factor here may be the difference between religiosity and spirituality.

There are certain prisoners who become rhetoric by their good deed. Its example is Gurung a quintessential Nepari, cheerful and fun loving whose life took a dramatic turn when Ghisingh launched the Gorkhaland agitation in 1986. He joined Gorkha Volunteers Cell, the police wing of the GNLF wet up on the lines of Hitler's SS. GVC cadres carried out kidnappings beheadings and bomb attacks against anti Gorkha element. Now at every occasion the public breaks into an impromptu jig with his supporters. He was the sufferer of poverty, deprivation and political violence born into a family of tea garden workers. The serious allegation led his ostracism. Now he is in Darjeeling living with great happiness.

A human being in a prison, particularly an under trial, is normally under high anxiety and stress. It is evident from research that yoga, meditation and religious practices have a positive impact to reduce stress and enhance mental health. Although, a few studies have been conducted on jail inmates (Eber, 1975; Khuruna, 1996), little has , however, been reported regarding the effect of Vipassana meditation on subjective well being of undertrials. Quality of Life is defined as the degree of excellence of one's life that contributes to satisfaction and happiness and benefits mental health. Quality of Life of an individual would be affected by a number of factors, particularly by the significant positive and negative life events. As stated by Milbrath (1979) 'Subjective studies of Quality of Life typically have shown that most people derive their greatest sense of Quality of Life from their home and family life and from the close supportive relationships they have with friends and colleagues'. Criminal Propensity is the notion of an underlying, or latent characteristic of all individuals - aggression, impulsiveness, self-control, or conditionality - that has a direct effect on a person's likelihood of committing criminal acts. Research findings suggest that people with Criminal Propensity score high on Neuroticism, Extroversion, Psychoticism and Lie tests. Therefore we poor in wellbeing.

In the past few decades, US policies have led to an unprecedented increase in the number of people behind bars. While more men than women are incarcerated, the rate of increase for women has been higher. Evidence of the negative impact of incarceration on the health of women of color have suggested strategies to reduce these adverse effects. Correctional policies contribute to disparities in health between White women and women of color, providing a public health rationale for policy change. Drug courts, treatment alternatives to street crime programs, and other mandatory treatment models have proliferated as policy makers strive to reduce reliance on costly custodial responses to drug-related crime. Although past research supports this approach, little is known about different forms of legal pressure used to compel treatment participation and their effects on client outcomes.

Results from a study of 161 offenders mandated from different criminal justice sources to attend long-term residential treatment are presented. Scores on an experimental perception of legal pressure measure co-varied predictably with the coercive strategies employed by the mandating sources, and multivariate analyses showed these scores to be relatively powerful, independent predictors of retention. Findings suggest providing information to clients about conditions and contingencies of treatment participation and convincing them they will be enforced are effective coercive approaches. There was less support for other forms of coercion—tight monitoring and use of severe penalties for failure (Young, 1989). *The study of Klein and Bahr (1992) evaluated a cognitive program at a state correctional institution. The program points to the benefits of cognitive skill development as inmates reflect on rejoining their families. Inmates are instructed on identifying problems, recognizing resources for solving problems, and thinking of alternative ways to respond as they apply principles of practical reasoning to everyday problems families experience. The development, implementation, and evaluation of the program and the implications for professionals were also evaluated and found useful for promoting their wellness.*

Specific roles for health professionals include becoming involved in alliances addressing alternatives to incarceration, creating programs that address the needs of women in correctional facilities, and identifying the pathways by which correctional policies damage health (Freudenberg , 2002).]t is undeniable that since approximately ninety percent of all inmates will one day be released, allowing prisoners to communicate with the outside world has important consequences: "Without such contact with society outside the prison walls, rehabilitation would be adversely affected, prison morale weakened, perhaps inviting riots and other forms of internal disorder, and the inmates' ability to readjust to the world outside the institution upon release would be markedly impaired. "Free speech in the prison context is also important since it is the only first-hand account we have as to how the penal system is run. Communication also plays a vital role in an inmate's access to the courts and his relations with counsel" (Holtz, 2002, p. 2). "Very often, inmates have used the Internet not only to look for pen-pals and counsel, but also as a way to convey the inner workings of the American prison system or to tell their stories in order to garner support from the outside. Others write of their personal growth while incarcerated or share their creative writing or art work" (Holtz, 2002, p. 3). "A clear and consistent emphasis on maximizing visitation and supporting contact with the outside world must be implemented, both to minimize the division between the norms of prison and those of the free-world, and to discourage dysfunctional social withdrawal that is difficult to reverse upon release" (Hany, 2002, p. 17). "The recommended policy guideline adopted by the Association of State Correctional Administrators on August 23, 1972, echoes the view that personal correspondence by prison inmates is a generally wholesome activity: 'Correspondence with members of an inmate's family, close friends, associates and organizations is beneficial to the morale of all confined persons and may form the basis for good adjustment in the institution and the community".

Growing prison populations in the U.S. are largely due to drug-related crime and drug abuse. Yet, relatively few inmates receive treatment, existing interventions tend to be short-term or non-clinical, and better methods are needed to match drug-involved inmates to level of care. Using data from the 1997 Survey of Inmates in State Correctional Facilities, a nationally representative sample of 14,285 inmates from 275 state prisons, we present a framework for estimating their levels of treatment need. The framework is drawn partly from the American Society of Addiction Medicine Patient Placement Criteria and other client matching protocols, incorporating drug use severity, drug-related behavioral consequences, and other social and health problems. The results indicate high levels of drug involvement, but considerable variation in severity/recency of use and health and social consequences. We estimate that one-third of male and half of female inmates need residential treatment, but that half of male and one-third of female inmates may need no treatment or short-term interventions. Treatment capacity in state prisons is quite inadequate relative to need, and improvements in assessment, treatment matching, and inmate incentives are needed to conserve scarce treatment resources and facilitate inmate access to different levels of care (Belenko, & Peugh, 2004).

Generally most of the criminal are of such kinds who are facing serious allegation but no body is standing in their support. According to Singh (2011) while there has been too much hype around the 17 Punjabi youths on death row in Sharjah with the Union government and political leaders of Punjab reaching out to them, after the case hit headlines back home, 45 more Punjabis are facing similar cases in UAE without help from any official. Social integration is linked to mental wellbeing among the general population and its relationship is not tested among prisoner. Jail is a unique institution which has direct relationship between jail institution and mental wellbeing. The study explored social integration and mental wellbeing among 198 male and female inmates incarcerated in jail. The study revealed that social and close relationship outside and inside the jail was associated with higher degree of psychological distress. Various attempt have been made to foster creativity of inmates lodged in jails of Kaithu and Kanda near Shimla to includes a theatre organization, Delhi Kalakaram/ Saroj Vashisth, a theatre coordinator of Delhi is conducting regular workshops for inmates since 1996 in different jails of Himachal Pradesh whose efforts is to bring out creativity. Warden of the jail are also putting efforts to re duce their mental symptoms through play and music (Dhaleta, 2011).

According to Sethi, N. (2011) food security is very important to the poor segments of the society and Sonia Ganghi the chairman of UPA govt. has did her best by sending such bill to the government. Divya (2011) There is nothing in such honorable killing...they are nothing but bar baric and shameful acts of murder committed by brutal, feudal-minded persons who deserve punishment (Supreme Court, 2011). The Supreme Court's condemnation has rattled Khap panchayats but they re rounding o it furiously and cannily accusing it of culpability for the very crime of which they have accused—honor killing. According to Dahiya there has been spurt in honour killings in Haryana in the last couple of year and the courts are getting it done. It reflect the typical mindset of the khap panchayat (Divya, 2011).

WELL-BEING OF THE MARGINALIZED PRISON INMATES

Approximately 450 million people are suffering from mental or behavioural disorder across the world as per the estimate of World Health Organization. Such disorders are prevalent among prisoner. According to US Department of Justice at mid year 2005 more than half of all prisons and jail inmates had a mental health problem including 705600 inmates in states prisons, 78,800 in Federal prison, and 479, 900 in local jails. These estimate represented 56% of State prisoners, 45% of Federal prisoners and 64% of jail inmates (Doris & Lauren, 2006). The survey of inmates in the states and Federal Correctional Facilicities , 2004 and the survey of inmates in the local jail 2002 included a modified structured clinical interview for DSM-IV. This survey indicates that the jail inmates suffered from major depression, mania and psychotic disorders. They reported depressed mood, decreased interest, or pleasure in activities, along with additional symptom of depression. The symptom of delusion and hallucination also were observed. Fellner (2006) also observed that the number of incarcerated men and women with severe mental illnesses has grown so tremendously in the last few decades that prisons may now be the largest mental health providers in the United States. Yet U. S.

Department are not well equipped for mentally ill prisoners. Prison conditions are hard on mental health in general

The prisons are often the scenes of brutality, violence and stress. A **prisoner** is anyone who is deprived of liberty against their will. This can be by confinement, captivity, or by forcible restraint. Prisoners are faced with incidence of violence and are always concerned for their safety. A long-term prisoner named Jack Abbott had stated "everyone is afraid. It is not an emotional or psychological fear. It is a practical matter. If you don't threaten someone at the very least, someone will threaten you...Many times you have to "prey" on someone, or you will be "preyed" on yourself" (Tosh, 1982:86). Prisons aim to cure criminals of crime however their record has not been encouraging. Instead prisons do more harm than good. The pains of jail confinement affect all prisoners in different ways. To begin with the prisoners need to withstand the entry shock by adapting quickly to prison life. The overcrowding, violence, solitude or conversely, lack of privacy and meaningful activities, isolation from social network, insecurity, and inadequate health service are some of the other factors. Prisoners are exposed to a new culture, which is very different from their own culture. Then they need to maintain outside links. For example, keeping in contact with family and friends becomes frustrating. While being in prison the prisoner must determine his/her way of passing the time since the hours appears endless (Tosh, 1982).

For some prisoners the major source of stress would include the loss of contact with family and friends outside the prison. There is also the fear of deterioration. There is lack of personal choice within the prison environment which many effect prisoners. After many years of being told what to do they may well lose the ability to think for themselves and make their own decisions and choices freely (Tosh, 1982). The purpose was to compare the social, emotional, and physical health characteristics of older and younger inmates of Canada. The results showed that when compared with younger inmates, the older men had more contacts with friends and family, more friends in prison, and experienced less stress than the younger men. Aside from problems with vision and hearing, their physical health did not appear to be any worse when compared to that of younger inmates (Elaine & Gallagher, 1990). The decade of the 1980s witnessed more than a doubling in the number of incarcerated individuals. Little is known about the psychological reactions of children whose parents are incarcerated, although a variety of behavioral disorders apparently related to separation, stigma, and deception of the child has been reported. The possibility of aggressive or antisocial behavior emerging in sons of incarcerated fathers has been mentioned as of particular concern in some reports. This article discusses salient themes in the literature on the reactions of children to parental incarceration, with an emphasis on boys' reactions to incarceration of their fathers. It critiques this literature and compares the findings with literature on the effects of separation in father absence related to other causes (for example, divorce, death, military service).

An Overview of Prison Reform

As per pre-history of the prison is concerned, it begins with Katherine Darton's in 10,000 BC where she reported that in prehistoric times there was, as far as historians can tell, no division between medicine, magic and religion. The disability social History timelines begins in 3,500 BC with an account of the fitting of an artificial limb the Rig-Veda.

The came S N Kothare and Sanjay A Pai's chapter on Evolution of Psychology and Psychiatry that discusses Ayurveda medicine which derives from the compendiums of Sushruta Samhita and Charaka date back to about the 6th century BC. In ancient **Greece and Rome** Larry Merkel's *History of Psychiatry,* available from his University of Virginia seminars begins with a discussion of pre-classical (Egyptian, Middle-Eastern, Judaic) influences on classical Greek and Roman theory and practice. Drama Therapy and Psychodrama History begins with plays of Sophocles in 404BC. Socrates recommends that "the offspring of the inferior, or of the better when they chance to be deformed, will be put away in some mysterious, unknown place, as they should be". Galens massive writing on medicine also appeared. In **1100** an asylum exclusively for sufferers from mental diseases at Mets" appeared. In **1188,** King Henry 2nd bought land next to Newgate in the City of London for a prison.

In **1201** St Nicholas Hospital in Carlisle claims to have been treating lepers in 1201 - to have passed to the City as a (general) hospital in 1477 - Some local historians link this forward to the opening of a fever hospital in 1809 and the Cumberland Infirmary in 1828. Frank Walsh in Union journal 1970. Noteworthy events in American Psychology begins with the founding of St Mary of Bethlem. In **1310** date was given for a German madhouse at Elbing near Danzig. De Praerogativa Regis, the Act giving the King custody of the lands of natural fools and wardship of the property of the insane, may have been drawn up between 1255 and 1290. In **1371** date was given for royal license to Robert Denton, chaplain, to use his own house in the parish of Berking, near the Tower of London, as a hospital "for the poor priests and for the men and women in the sad city who suddenly fall into a frenzy and lose their memory, who were to reside there until cured; with an oratory to the said hospital to the invocation of the Blessed Virgin Mary".

In the year 1377, the religious priory of St Mary of Bethlem, in London, was confiscated by King Edward 3rd in 1375, and used for lunatics from 1377. (Jones 1972 p.12). In **1403/1404** it had just six insane patients and three who were sane. (Scull 1972 p.19). This old **Bedlam** was a small institution (on a site south of what is now Liverpool Street Station), even in the 17th century when it had about 30 patients. Its showy replacement, the Moorfields Bedlam, opened in 1676.The priory of St Mary of Bethlem was founded in 1247 as a priory in Bishopsgate Street, for the order of St. Mary of Bethlehem, by Simon Fitz Mary, an Alderman and Sheriff of London. During **1403** *Report of a Visitation which had enquired into the deplorable state of affairs at Bethlem Hospital* (Michael Warren) appeared. In **1464** Michel Foucault suggests that the publication of Brant's illustrated narrative poem marks a point in European culture where a dialogue between reason and unreason became central. In **1495** Syphilis, possibly introduced from the new world, broke out amongst troops in Italy and rapidly spread across Europe, reaching England and Holland in 1496. It reached India in 1498. In 1500 there was an epidemic of syphilis across Europe and in 1505 it reached China. In 1500 Zilboorg writes of some sixteenth century writers as "*The first psychiatric revolution*" and (Ackernecht 1968, p-) writes of the "magnificent developments of psychiatry" in the 16th century.

In the year 1518, King Henry 8th, on the advice of his court physician, founded the Royal College of Physicians (London) to control who practiced as a physician in London and so protect the public from quacks. In the year 1520, a small book by Paracelsus, written about 1520 and published 1567 was called "Diseases which lead to a Loss of Reason" In the year **1536,** First Act of the Dissolution of the Monasteries. Although the religious foundations were closed, any "hospital" (refuge for the homeless poor) attached might continue. In the year **1547** King Henry 8th signed a document giving Bethlem Hospital and St Bartholomew to the City of London. The name "St Bartholomew" being changed to "the House of the Poor in West SmithfielThe City of London unsuccessfully petitioned the King to give them five hospitals plus their endowments. The hospitals included Bethlem, St Bartholomew and St Thomas. From 1557, Bethlem was managed by the governors of the Bridewell House of Correction (established 1550). The governors were chosen by the City of London. Bethlem was controlled by the City of London until it was transferred to the National Health Service in 1948. Further Philip Barrough during 1560 to 1590 method of physic, containing the causes, signs, and cures of inward diseases in man's body from the head to the foot come forth.

In the year 1592 An account of a trial for conspiracy to kill the King, written by Richard Cosin, contains discussion and definitions of the terms applying to the various "degrees" of insanity. In 1611 Authorised (King James) version of the Bible. The bible was a major source for ideas about virtually everything in the 17th century, and later. In her *Notes of the history of mental health care* (on the *Mind*website), Katherine Darton outlines some of its influences in her consideration of the Jewish tradition. (Scroll down from 2,000BC). In **1615,** *Giles Earle His Booke,* a manuscript collection of lyrics in the British Museum, contains the first known written version of the English Folk lyrics "Tom o'

Bedlam's Song" appeared. Further Winter 1651 George Fox's vision of blood in the streets of Lichfield appeared. From November 1660 (arrested) to 1672, John Bunyan, a Baptist preacher, was imprisoned almost continuously in Bedford Gaol for preaching outside the established church. In **1690** an essay concerning understanding, John Locke said there is a degree of madness in almost everyone concept appeared. In eighteenth Century Asylums appeared and English asylums in the eighteenth century were small and they were not run by the state. The best known and the largest was *Bedlam* or *Bethlem* in the city of London. This had 130 patients in 1704. Ackernecht (1968, p.) argues that psychiatry "reached the status of an independent science" in the eighteenth century. But not due to "developments in medicine but to the philosophy of enlightenment which pervaded the whole century". Reasons were: belief in "possession by evil spirits" came to be regarded as "superstition". Reason was the highest good for the philosophers, so they sympathized especially with those who lost their reason. He argues that: "since the concept of the immortal soul was of no importance in this philosophical system, mental disorders could be viewed as disorders of the mortal brain or thinking apparatus and as such could now at last be studied on a scientific basis.

In the year 1700, David Irish in his madhouse near Guildford, Surrey, claimed to cure by good food and comfort, and would care for those who were not curable for life. In the 1714 Vagrancy Act is thought to have been the first English statute to provide specifically for the detention of lunatics, but Blackstone argues that it was based on common law. Under **1744** Vagrancy Act, construction of naval hospitals at Gosport (Haslar), Plymouth and Chatham authorised. The Haslar hospital was built between 1745 and 1761. "The hospital catered for a full range of illnesses and included wards for medical, surgical, fever, flux, small pox, consumptive, scorbutic and recovery as well as lunatics. In the late 19th and early 20th centuries, Haslar was one of the most important naval hospitals in the country. It became the main lunatic asylum for the navy as well as providing for infectious diseases between 1898-1902" (PAPHE external link) - The navy placed lunatic patients in Hoxton House, at least until 1818, but also had insane patientsat Haslar.

The use of solitary confinement can be traced far back but became common with the rise of the modern penitentiary during the first half of the nineteenth century. The practice of isolating individual prisoners has changed significantly since then but has remained a feature of Western prison systems. A debate about the effects of solitary confinement was largely settled early in the twentieth century, when the use of large-scale solitary confinement appeared to be on the way out in most Western prison systems. Discussions on the effects of solitary confinement resurfaced in the 1950s when sensory deprivation and perceptual deprivation studies were carried out partly in reaction to stories of brainwashing of U.S. prisoners of war (POWs) during the Korean War. These studies waned by the early 1970s. During the 1980s solitary confinement again regained topicality when supermax prisons caused an explosion in the use of solitary confinement. Probably much less known, pretrial solitary confinement has for many years been an integral part of Scandinavian prison practice and has in recent decades been considered a significant problem (Smith, 2006)

Social Wellbeing among the Prisoner

The prisoner or jail inmates are prone to various social stressors. There is a differential effects of imprisonment on the inmate's family—the hidden familial costs of imprisonment. Lengthy interviews were conducted with adult members of inmates' families in a sample drawn from a maximum security prison, and 46 of these are analyzed here. Problems with income, employment and child care were severe, pervasive, and subject to only minor changes over time. These affected spouses more than other family members. Deterioration of relationships with family and friends were also noted by many, showing a greater tendency for adjustment over time. Some of the strongest perceived effects of imprisonment on the family members were emotional, including problems of expressing and dealing with feelings, and problems of experiencing enjoyment in life. Social ties complicate the punitive consequences of incarceration, which stands in tension with a system of justice that is oriented to the individual rule violator. The drawings of Todd, a prisoner confined in a Pennsylvania "close-security" or "super maximum" prison, tell a story—one that graphically portrays the devastating effects of a prison on the mental health of its inmates.

This study examines gender similarities and differences in background characteristics, the effectiveness of treatment, and the predictors of post-release outcomes among incarcerated drug-using offenders. The sample of 1,842 male and 473 female treatment and comparison subjects came from a multi-site evaluation of prison-based substance abuse treatment programs. Three-year follow-up data for recidivism and post-release drug use were analyzed using survival analysis methods. Despite the greater number of life problems among women than men, women had lower three-year recidivism rates and rates of post-release drug use than did men. For both men and women, treated subjects had longer survival times than those who were not treated. There were both similarities and differences with respect to gender and the other predictors of the two post-release outcomes. Differences in background characteristics and in factors related to post-release outcomes for men and women suggest the plausibility of gender-specific paths in the recovery process **(Pelissier, Camp, Gaes, Saylor & Rhodes, 2003).**

The framework is drawn partly from the American Society of Addiction Medicine Patient Placement Criteria and other client matching protocols, incorporating drug use severity, drug-related behavioral consequences, and other social and health problems. The results indicate high levels of drug involvement, but considerable variation in severity/recency of use and health and social consequences. The researchers estimated that one-third of male and half of female inmates need residential treatment, but that half of male and one-third of female inmates may need no treatment or short-term interventions. Treatment capacity in state prisons is quite inadequate relative to need, and improvements in assessment, treatment matching, and inmate incentives are needed to conserve scarce treatment resources and facilitate inmate access to different levels of care (Belenko, & Peugh, 2004).

Growing prison populations in the U.S. are largely due to drug-related crime and drug abuse. Yet, relatively few inmates receive treatment, existing interventions tend to be short-term or non-clinical, and better methods are needed to match drug-involved inmates to level of care. Using data from the 1997 Survey of Inmates in State Correctional

Facilities, a nationally representative sample of 14,285 inmates from 275 state prisons, we present a framework for estimating their levels of treatment need. The framework is drawn partly from the American Society of Addiction Medicine Patient Placement Criteria and other client matching protocols, incorporating drug use severity, drug-related behavioral consequences, and other social and health problems. The results indicate high levels of drug involvement, but considerable variation in severity/recency of use and health and social consequences. Result revealed that one-third of male and half of female inmates need residential treatment, but that half of male and one-third of female inmates may need no treatment or short-term interventions. Treatment capacity in state prisons is quite inadequate relative to need, and improvements in assessment, treatment matching, and inmate incentives are needed to conserve scarce treatment resources and facilitate inmate access to different levels of care (Belenko, & Peugh, 2004).

The study **(Otto, & Dalbert, 2005)** focuses on the relationship between belief in a just world (BJW) and the feelings and behavior of young prisoners, taking into account their familial background and criminal career. Our sample comprised 66 young male prisoners from a German detention center. Regression analysis provided evidence for the three functions of the BJW: (a) High believers showed more justice motive-congruent behavior, e.g., they revealed fewer disciplinary problems during imprisonment. (b) High believers interpreted events in their life as more just, e.g., they perceived their legal proceedings as more just; consequently, they reported more feelings of guilt, and they were better able to cope with their anger. (c) High believers revealed more trust in their future, e.g., they were more confident that they would be able to achieve their personal goals. Overall, the BJW seems to be an important resource capable of improving young prisoners' rehabilitation prospects.

The crime is a major activity in the US, with implications for poverty and the allocation of public and private resources. The economics of crime focuses on the effect of incentives on criminal behavior; the way decisions interact in a market setting; and the use of a benefit-cost framework to assess alternative strategies to reduce crime. This essay shows that most empirical evidence supports the role of incentives in the criminal decision: legitimate labor market experiences, sanctions including incarceration, and the risk of apprehension all influence decisions to engage in crime. By putting crime into a market setting, economic analysis highlights the difficulty of reducing crime through incapacitation: when the elasticity of supply to crime is high, one criminal replaces another in the market; and thus the importance of deterring crime by altering behavior. Most analyses show that "crime pays" in the sense of offering higher wages than legitimate work, presumably in part to offset the risk of apprehension. But some important facts about crime — long term trend increases and decreases; the geographic concentration of crime; the preponderance of men and the young in crime — seem to go beyond basic economic analysis (Freeman, 2007).

To analyze the relationships between mental health and employment commitment among prisoners and the long-term unemployed (LTU) trying to return to work. Fifty-two of 62 male inmates of a semi-open prison (Givenich Penitentiary Centre, the only such unit in Luxembourg), and 69 LTU registered at the Luxembourg Employment Administration completed a questionnaire exploring: 1) mental health (measured by means of scales GHQ12 and

CES-D); 2) employment commitment; 3) availability of a support network, self-esteem, empowerment; and 4) socio-demographic characteristics. Results revealed that Compared with LTU, inmates were younger, more had work experience (54.9% *vs* 26.1%), and more were educated to only a low level (71.1% *vs* 58.0%). The link between employment commitment and mental health in the LTU was the opposite of that seen among the prisoners: the more significant the perceived importance of employment, the worse the mental health (GHQ12 $p = 0.003$; CES-D $p < 0.001$) of the LTU; in contrast, among prisoners, the GHQ12 showed that the greater the perceived value of work, the lower the psychic distress ($p = 0.012$). Greater empowerment was associated with less depression in both populations. The education levels of people who did not reach the end of secondary school, whether inmates or LTU, were negatively linked with their mental equilibrium. The two groups clearly need professional support. Future research should further investigate the link between different forms of professional help and mental health. Randomized controlled trials could be carried out in both groups, with interventions to improve work commitment for prisoners and to help with getting a job for LTU. For those LTU who value employment but cannot find it, the best help may be psychological support (Meyers, Le Behan, & Housemand, 2008).

The U.S. Supreme Court first recognized the problem of prison rape in *Farmer v. Brennan* (1994), unanimously holding that the Eighth Amendment's ban against cruel and unusual punishment would be violated if prison guards acted with "deliberate indifference" and "ignored a substantial risk of serious harm" to the inmate, noting that sexual abuse is "not part of the penalty that criminal offenders pay for their offenses against society" (511 U.S. 825). The problem became so prevalent and alarming that in 2003 the U.S. Congress passed the Prison Rape Elimination Act (PREA) to identify, prevent, prosecute, and respond to prison sexual violence in correctional facilities (Tess & Carl, 2010).

The jail inmates sometime die inside the jail that put question mark on the facilities being provided by the authorities inside the prison. Its example is Ferozepur Jail where death of other inmates has once again spotlighted due to insufficient medical care. This was the eighth death in this jail. The elite and high income class is mostly associated behind heinous crime. The rape new of dalits girls and married one are the common one those really are the sufferer (Kamal, 2010). Several inmates after inquiring reported that medical facilities here are not significant (Sharma, 2011). Beside this like in Haryana state if dalits even practice marriage with upper caste girls then she is killed for honor again raise a question towards the Indian constitution Amendment and removal of caste system. According to Mahapatra (2011) there is perhaps a doubt whether the socially beneficial tool are turning into legal terrorism? Misuse of Section 498A seems to have resulted in innocents get ting punished.

The supreme Court faced an identical dilemma just last year, relating to dowry harassment cases under Section 498 A of Indian Penal Code "It is a common knowledge that unfortunately, matrimonial litigation is rapidly increasing in our country.....this clearly demonstrates discontent and unrest in the family life of a large number of people of society. On the one hand the law was the social legislation giving women protection against harassment at the hands of the husband and his relatives who demand more and more dowry the court was concerned with the rapid rise in such cases. But on the other hand, it was aware that a large number of such complaints are not bonafide and are filed with oblique motive. It admitted that to find out the truth is Herculean task and the tendency of implicating husband and all his immediate relations is also not uncommon. Thus it is difficult to ascertain the real truth. After death and loss of any kind takes place inside the jail, it took much time to compensate the affected people. Its example is Mumbai Jail where court vicariously liable for the death and order Rs. Four lakh to pay. Criticizing the delay in the system that deprived the family of financial help for a long time, a division bench of Justice Joshi and Bakre directed the government to provide such compensation to the victims that perhaps is a good sign but is a late gesture (Thomas, 2011). It seems that the Supreme Court is playing pivotal role for compensating the deprived people of our country. Families of those who died during clinical trials last year have finally been paid compensation by the pharmaceutical companies which owned these drugs (Sinha, 2011).There are certain oldest prisoner who walks freely now prefer non-vegetarian food, and prayed every day (Srivastava, 2011).

According to Sen (2011) good inmates are those who get open air in jail. Jharkhand has become first even "open air" jail that is ready to accommodate 100 convicted prisoners. The open air facilities have been provided to those inmates in jail who has good record during their jail term. According to Kumar (2011) there are also certain prisoner who do not want to go to their country or home because they find insecurity to return there after a long period of time. Further they find helplessness to resettle their due to social pressure they experience there. Its example is the prisoner Sindh, Karachi who when burnt Pakistani flag at Jantar Mantar Delhi now fear to go to Pakistant due to the retaliation of the government. Level of pollution in the jail is continuously increasing. As a result the prisoner suffers from health problem such as blood pressure and blockage in arteries as well as respiratory, itching in eyes and skin irritations. Not only the prisoner but also the staffs attending them are also suffering. The cause of pollution is establishment of industries ear jail, brick kilns, quarries crushing stones, unchecked growth of chemical factories, restricted construction activity and frequent fires in the vicinity . A cloud of dust and smoke hanging over the area, air and ater pollution and non functional sewerage plants are responsible for it (Thomas, 2011).

Psychological Wellbeing among Jail Inmates

A competitive survey was carried out by Elaine and Gallagher (1990) in three medium-security Federal prisons in British Columbia, Canada compared the social, emotional, and physical health characteristics of older and younger inmates of Canada. The results showed that when compared with younger inmates, the older men had more contacts with friends and family, more friends in prison, and experienced less stress than the younger men. Aside from problems with vision and hearing, their physical health did not appear to be any worse when compared to that of younger inmates. This was surprising in light of age-related declines in health found in both the general population and in previous research with inmates. The study uncovered several unique problems concerning older prisoners, and a number of policy and program recommendations were generated.

The study of Ivanoff, Smyth, Grochowski, Jang, and Klein (1992) examined the relationships between means–ends problem solving and suicidality among adult male prison inmates in light of new evidence based on inpatient and college student populations suggesting that state, rather than trait, vulnerabilities may be responsible for problem-solving deficits and differences. Using the Means–Ends Problem-Solving Procedure with 93 state prison inmates, it was found that among inmates with a history of parasuicide, current suicidality did not affect problem-solving performance. Among nonsuicidal inmates, parasuicide history had no effect on problem solving or affect–suicidality measures. Although these results support new research suggesting that trait problem-solving deficits are not causally linked to suicidality, they raise questions about the potentially unique relationships among suicidality, problem solving, depression, and hopelessness in incarcerated populations. Behavioral or emotional disorder associated with paternal incarceration probably is related mainly to associate factors such as the meaning of the incarceration to the child, the remaining caretaker's psychological characteristics and psychopathology, the parenting relationship between the caretaker and the child, and the coping capacities and resources of the family, rather than to the separation itself. Recommendations for further research in this increasingly important field are provided (Gabel, 1994).

So it is very important to prepare the soldier mentally to face any type of the situation. Krishna (1993) reports the presence of neuroticism, anxiety, extraversion, and morality guilt among adolescents who are high on delinquent behaviour. Accordign to Ogloff, Roech, & Hart (1994), united state has the largest per capita prison population in the industrialized world. Here 6.5% to 10% inmates suffer from serious mental illness and 15% to 40% suffer from moderate mental illnesses. She says that there are positive relationships between delinquent behaviour and these personality factors. The study of Osofsky (1996) reveals the presence of certain psychological or personality factors exhibited by prisoners and the importance of these in creating more stress within the prisons. These factors are neuroticism, anxiety, aggression, hostility, and guilt. It has been found that severe psychopathological emotion is higher in adolescents exhibiting higher degree of offences. Ahmad (1988) has also reported that mediators show overall better adjustment and personality organization than non-mediators. About **21-45% of criminals in prison have ADHD** 15 clinical studies from peer reviewed journals show that I list below. Only 8% of children and 5% of adults have ADHD.

In their study **Jon and Gisli (1996) observed** psychological characteristics of 62 prison inmates, who claimed to have made a 'false confession' during a police interview, were compared with those of other inmates. The personality tests administered were the Eysenck Personality Questionnaire ,the Gough Socialisation Scale, the Gudjonsson Suggestibility Scale. The Gudjonsson Compliance Scale, and the self-deception and other-deception questionnaires of Sackeim and Gur. In addition, Ravens' Standard Progressive Matrices were administered. The results showed that the 'false confessors' were significantly more antisocial in their personality, compliant and emotionally labile, than the other inmates and had lower self-deception and other-deception scores. A subgroup of coerced-internalized 'false confessors' was identified and they differed from the others in terms of elevated suggestibility and confabulation scores as measured by the GSS 1.

In a parallel study of 108 juvenile offenders it was found that none of them reported having made a 'false confession', in contrast to 12% of the prison inmates. Overall, the results indicate that 'false confession' in Iceland may be largely confined to persistent offenders who possess antisocial personality characteristics (**Jon & Gisli, 1996**). Conventional wisdom has regarded low self-esteem as an important cause of violence, but the opposite view is theoretically viable. An interdisciplinary review of evidence about aggression, crime, and violence contradicted the view that low self-esteem is an important cause. Instead, violence appears to be most commonly a result of threatened egotism—that is, highly favorable views of self that are disputed by some person or circumstance. Inflated, unstable, or tentative beliefs in the self's superiority may be most prone to encountering threats and hence to causing violence. The mediating process may involve directing anger outward as a way of avoiding a downward revision of the self-concept (Baumeister, Smart, L. & Boden, 1996).

The reviews and critiques Mitchell (1999) on major studies had shown possible relationship between mental disorder and violent offending. Such studies have either examined a) the prevalence of violence in the mentally disordered, or b) the prevalence of mental disorder amongst the violent. Indices of violence used include a) arrest rate, b) conviction rate and c) prevalence of violent behaviour in psychiatric inpatients, representative community samples, and birth cohorts. It is concluded that although the mentally disordered do face an elevated risk of violent behaviour/offending, this risk is largely confined to those suffering from severe or long term psychotic symptoms, or psychopathic disorders. Incarcerated women have numerous physical, social, and emotional health care needs, including specific needs related to their expressions of sexuality while in prison. This report describes the results of a participatory action research study with incarcerated women utilizing critical hermeneutic data analysis techniques. While the public's view of sexuality between incarcerated women borders on the prurient and profane, this study suggests that women in prison continue to be sexual beings who come to "participate" in love and sex with one another based on their need for relationship and friendship. It is suggested that prison bureaucracies define women through a sexual lens, dually grounding their identities in the crimes they were sent to prison for and the perceived crimes of their sexuality. The need for nursing involvement and intervention with this marginalized and stigmatized population is discussed (Maeve, M. Katherine R. N, 1999). The existence of prison rape in male prisons in the United States is a serious problem with pervasive and devastating consequences (Dumond, 2000).

Few adolescents who experience significant psychological distress seek professional psychological help, a finding particularly pertinent in New Zealand which has one of the highest youth suicide rates in the world. In the present study, 221 New Zealand high school students completed a questionnaire which examined the relationship between a variety of approach and avoidance factors associated with professional psychological help-seeking. Suicidal ideation, attitudes, psychological distress, treatment fears, gender and prior help-seeking were significant predictors accounting for approximately 23% of students', self-rated help-seeking intentions. Contrary to expectations, higher levels of suicidal ideation led to lower levels of help-seeking intentions for suicidal thoughts. This findings is discussed in relation to the process of help-negation which has been identified in clinical suicidal samples. The implications of these findings for interventions that increase appropriate professional psychological help-seeking in adolescents are also discussed **(Pamela & Frank, 2000).**

One hundred and fifty felony drug offenders diverted from prison to community-based, residential drug treatment alternative to prison program completed a comprehensive interview as part of a longitudinal study. Treatment completion predictors were sought examining intake data (demographics, family, social, employment, medical, psychological, criminal, sexual behavior, drug use and treatment histories). Logistic regression results found completers had more social conformity and close friends, and less need for employment counseling, felony drug convictions, drug dealing income, and unprotected sex than dropouts. Completers were also less likely to encounter recent problems with significant other, have a psychiatric history, experience gunshot or stabbing, and commenced heroin use at older ages than dropouts. However, completers reported higher alcohol use than counterparts. Further analyses explored subcategory models: "life choice" (substance use, criminal and sexual behavior), static (background and dispositional), and dynamic situational influences (**Michelle & Steven, 2000**).

The prison environment has the psychological consequences such as overcrowding (Haney, 2001). Evidence of the negative impact of incarceration on the health of women of color suggests strategies to reduce these adverse effects. Correctional policies contribute to disparities in health between white women and women of color, providing a public health rationale for policy change. Specific roles for health professionals include becoming involved in alliances addressing alternatives to incarceration, creating programs that address the needs of women in correctional facilities, and identifying the pathways by which correctional policies damage health (Freudenberg, 2002). According to Baumeister, Catanese, and Wallace (2002) men's efforts to force women to engage in unwanted sexual activity can be explained by a combination of reactance theory and narcissism. Reactance theory also suggests that deprivation of specific sexual options will cause men to desire them more, to try to reclaim them by forcing sex and by aggressing against the woman who has refused them, and assorted findings support this analysis. Narcissism is proposed to moderate the link, especially because coercion is relatively rare in response to sexual refusals. Evidence about sexually coercive men supports the narcissism hypothesis, such as by showing self-serving cognitive distortions, an excessive concern with being admired, an inflated sense of entitlement, selectively low empathy, and a broadly exploitative approach to heterosexual relations.

Each year, more than 10 million people enter US jails, most returning home within a few weeks. Because jails concentrate people with infectious and chronic diseases, substance abuse, and mental health problems, and reentry policies often exacerbate these problems, the experiences of people leaving jail may contribute to health inequities in the low-income communities to which they return. Psychosocially vulnerable prisoners under stressful conditions of confinement are ill prepared to cope and at risk for developing suicide intention. The present study examined the relationships of depression, hopelessness, and reasons for living, mental health problem history, suicide attempt lethality history, and stressful segregation housing with suicide ideation in 134 prisoners. Prisoners housed in segregation were found to have significantly higher levels of depression and suicide ideation than prisoners in general population. A hierarchical regression model of suicide ideation found significant interactions between mental health problem history, suicide attempt lethality history, and hopelessness with anticipated segregation stress, independent of depressed mood. Results of the study are discussed in terms of the stress-vulnerability model, various methodological limitations, and future research (Bonner, 2006).

Given the physical and psychological trauma many victims experience, it is particularly important to sharply reduce sexual assault in prisons. Such assaults are not only criminal in nature but also are crimes that take place in a facility created, funded, and operated by the state, which bears the responsibility of keeping its wards from predictable and preventable harm (Wolff, Shi, Blitz, & Siegel, 2007). In the prison system there are high percentages of mentally disordered offenders, often with co-morbid psychiatric disorders. In addition, the setting and ward atmosphere in prisons are not designed to care for psychiatric patients and places a burden on vulnerable inmates. Management of care of the different subgroups in the prison population is therefore necessary on four different levels, from basic health care to forensic psychiatric treatment. A fifth level is forensic care, which is directed towards re-integration into the community rather than treatment at the symptom-level of disease. Continuity of care, evidence-based care and coercion within mental health care are discussed with regard to their forensic ethical meaning towards both patients and the community. Drawing on the Dutch situation a case is made for the promotion of voluntary treatment for those prisoners with mental health problems to aid their rehabilitation and to lessen the risk to the public at large (Van Marle, 2007).

The setting and ward atmosphere in prisons are not designed to care for psychiatric patients and places a burden on vulnerable inmates. Management of care of the different subgroups in the prison population is therefore necessary on four different levels, from basic health care to forensic psychiatric treatment. A fifth level is forensic care, which is directed towards re-integration into the community rather than treatment at the symptom-level of disease. Continuity of care, evidence-based care and coercion within mental health care are discussed with regard to their forensic ethical meaning towards both patients and the community. Drawing on the Dutch situation a case is made for the promotion of voluntary treatment for those prisoners with mental health problems to aid their rehabilitation and to lessen the risk to the public at large (van Marle, 2007).

Increasing numbers of mentally ill offenders without concomitant increases in mental health staffing, issues of service utilization and priority populations have become paramount. The aim of **Morgan, Steffan, Shaw, and Wilson, (2007) study was** to expand upon previous studies of correctional mental health service utilization by examining the underlying factor structure of problems for which inmates may seek services and by examining the barriers hindering inmates' willingness to seek services. Additionally, this study investigated whether previous treatment experiences had an impact on inmates' willingness to seek services or barriers to seeking services. Participants consisted of 418 incarcerated adult males who voluntarily participated in this study and were from three security levels (that is, reception and diagnostic unit, minimum security, and maximum security). Participants completed a three-page survey inquiring about use of mental health services before and during incarceration. *Results based on f*actor analysis indicated five types of problems for which inmates may request mental health services, physical health concerns, negative affect, interpersonal relationships, and institutional relations) and four types of potential barriers to service utilization (self-preservation concerns, procedural concerns, self-reliance, and professional service provider concerns). Inmates with no prior history of mental health service use were more likely to present with self-preservation concerns about services and present as self-reliant. Results of this study provide an overarching conceptualization of barriers hindering inmates' willingness to seek services, as well as the types of problems for which they are likely to seek services. Also, inmates' mental health treatment history affected their current perceptions of services.

A study of Steadman, Fred , Osher, Robbins, Case, and Steven Samuels (2009) estimated current prevalence rates of serious mental illness among adult male and female inmates in five jails during two time periods (four jails in each period). During two data collection phases (2002–2003 and 2005–2006), recently admitted inmates at two jails in Maryland and three jails in New York were selected to receive the Structured Clinical Interview for DSM-IV (SCID). Selection was based on systematic sampling of data from a brief screen for symptoms of mental illness that was used at admission for all inmates. The SCID was administered to a total of 822 inmates—358 during phase I and 464 during phase II. To determine the current (past-month) prevalence of serious mental illness (defined as major depressive disorder; depressive disorder not otherwise specified; bipolar disorder I, II, and not otherwise specified; schizophrenia spectrum disorder; schizoaffective disorder; schizophrenic form disorder; brief psychotic disorder; delusional disorder; and psychotic disorder not otherwise specified), interview data were weighted against strata constructed from the screening samples for male and female inmates by jail and study phase. Result revealed that the rate of current serious mental illness for male inmates was 14.5% (asymmetric 95% confidence interval [CI]=11.0%–

18.9%) and for female inmates it was 31.0% (asymmetric CI=21.7%–42.1%). The estimates in this study have profound implications in terms of resource allocation for treatment in jails and in community-based settings for individuals with mental illness who are involved in the justice system. The mental health correlates of sexual trauma are complex, and for many victims, the pattern of symptom development may not be fully encompassed by existing official diagnostic categories (Cockrum, 2009).

Suicide is one of the leading causes of prisoners' death, and, as a group, inmates have higher suicide rates than their community counterparts. However, the prevalence of suicidal ideation among Chinese inmates and their risks of suicide remain unknown. Aims: This study investigates the prevalence of suicidal ideation among Chinese incarcerated populations and its risk and protective factors. Methods: We randomly selected three types of prisoners (N = 690) and a sample of college students (n = 248) from four correction facilities and a police academy in China. A self-report questionnaire with standardized scales translated from the West was administered to the samples. Results: The prevalence of suicidal ideation among Chinese inmates is similar to that found among inmates in the West. The correlates of suicidal ideation for the Chinese inmates are also in line with what has been found in general populations. The study suggest that Chinese inmates have the same risk factors and protective factors for suicidal ideation as the general population in China. Suicide prevention in Chinese prisons can learn from and inform suicide prevention work for the general population (Zhang, Grabiner, Virginia, Zhou & Li, 2010). According to Sinha (2011) by 2020, 15% of world ailments will be due to mental disorders.

Well-being of the Marginalized Prison Inmates

The effects of imprisonment on a woman's self concept are compounded by the pre-existing deterioration by other forms of deprivation and societal marginalization. Female long term prisoners tend to be poorly educated and unemployed. Many have been victims of sexual and physical abuse. Their life circumstances, along with feelings of guilt, fear, anxiety, alienation and confusion which are aggravated when they are arrested and sentenced by the justice system, combine to produce a group of women with extraordinarily low self esteem. Low self esteem reduces a woman's ability to cope and increases the self destructive behaviours so prevalent among federally sentenced women. Self injurious behaviour is common among female long term inmates, many of whom slash themselves with razors, knives, or other sharp items. A 1989 survey of women in the Kingston Penitentiary for Women (P4W), 59% indicated that they have engaged in self injurious behaviours; and of these, 92% indicated that they has slashed themselves (Heney, 1990). In addition to contributing to the incidence of self injurious behaviour, a lack of self esteem also reduces a person's ability to plan for the future, take responsibility for her actions and to believe that she can make meaningful choices (Task Force on Federally Sentenced Women, 1990).

For many female long term inmates, dependencies on men, alcohol or drugs and/or on state financial assistance have greatly reduced their ability to make choices. Upon being incarcerated, federally sentenced women continued to have very limited choices. Few training and support programs are available, work options are severely limited and choices over place of incarceration carry heavy consequences.

Canadian correctional institutions have been criticized for their tendency to encourage dependent and child-like behaviour among female inmates. Many female inmates have described the seemingly arbitrary rules and regulations of incarceration and how these rules humiliated them and contributed to their sense of powerlessness. Denied privacy, quiet and dignity, many women reported feeling they had no rights or control, which led to an overwhelming sense of hopelessness and lack of motivation (Task Force on Federally Sentenced Women, 1990).

Gender-based violence - including rape, domestic violence, mutilation, murder, and sexual abuse — is a profound health problem for women across the globe. Although a significant cause of female morbidity and mortality, gender violence is almost never seen as a public health issue. Recent World Bank estimates of the global burden of disease indicate that in Established Market Economies, gender-based victimization is responsible for one out of every five healthy days of life lost to women of reproductive age. On a per capita basis, the health burden imposed by rape and domestic violence is roughly equivalent in both the industrial and developing world, but because the overall burden is so much greater in the developing world, the percentage attributable to gender-based victimization is smaller (World Bank, 1993). Nonetheless, on a global basis, the health burden among reproductive age women of gender-based victimization is comparable to that posed by other conditions already high on the world agenda . More appropriately the gender-based violence includes a host of harmful behaviors that are directed at women and girls because of their sex, including wife abuse, sexual assault, dowry-related murder, marital rape, selective malnourishment of female children, forced prostitution, female genital mutilation, and sexual abuse of female children. Specifically, violence against women includes any act of verbal or physical force, coercion or life-threatening deprivation, directed at an individual woman or girl that causes physical or psychological harm, humiliation, or arbitrary deprivation of liberty and that perpetuates female subordination (Heise et al., 1994).

Because women make up only 2.6% of the long term inmate population in Canada (women account for 179 of the 7,004 long term prisoners)(Task Force on Long - Term Offenders, 1998), the tendency until very recently has been to focus research primarily on the behaviour and needs of male inmates. In Canada, the emergence of the feminist perspective in the understanding of social inequity and the recognition of the special needs of female offenders led to the appointment of the Task Force on Federally Sentenced Women in 1990. This group of corrections officials, other representatives of the justice system and community members studied the experiences of female federal inmates and made recommendations as to how the CSC could better manage these women. The Task Force produced a report that was a cooperative effort on the part of the Correctional Service of Canada, the Elizabeth Fry Society and Aboriginal women's groups. In their report, some insights into the effects of both long and short term incarceration on female inmates were provided.

The 1990 Task Force collected experiences of federally sentenced women, some of whom were serving long term sentences, and included them in the report. The women who contributed to the Task Force survey wrote about the effect that incarceration had on them. Many of these women spoke of the intense pain and anxiety caused by the separation from their children and of their sense of powerlessness when their children were placed in foster homes. The inmates who chose to remain in their home provinces under exchange of service agreements had done so primarily to maintain regular contact with their children. The cost of transportation, the willingness of foster parents to facilitate visits and the cost of telephone calls were all factors affecting the inmates' ability to maintain contact. When the Task Force report was submitted, women were not allowed to have their children with them in prison, but today, several of the newly established federal institutions for women provide mother-child programs that allow women who meet specific criteria to live with their children in the institution.

The Correctional Service of Canada completed its review of the Task Force on Federally Sentenced Women in 1991. It gave its support to eight recommendations aimed at improving conditions at the Prison for Women and began implementing them. The recommendations included: - changing the policy for transfers to the Prison; improving counselling services for victims of sexual abuse and family violence; preventing self injurious behaviour; ensuring the daily presence of an Aboriginal elder; helping inmates maintain and strengthen family and cultural ties; and promoting outdoor exercise and walks through improvements to the physical facilities (Solicitor General of Canada, 1991). The Task Force recommended stressed on to 24 hours talking circles for mediating the conflicts and increasing community participation (Green, 1997).

This article examines the subjective experiences of previously imprisoned women. Their retrospective narratives of prison life reveal overt behavioral and underlying structural tensions that create an atmosphere of fear and violence. Furthermore, attitudes of indifference between inmates and correctional staff often contribute to fostering an environment of neglect. The study, based on in-depth interviews with fifty-four female subjects, describes and analyzes several aspects of the socialization process for inmates as related by women on parole. The research shows that the "pains of imprisonment" for women are suffered to a greater degree than previously acknowledged. Prison for these women is a social world filled with anxiety and, perhaps, represents a punishment well beyond what the law intended **(Pogrebin & Dodge, 2001).**

In the past few decades, US policies have led to an unprecedented increase in the number of people behind bars. While more men than women are incarcerated, the rate of increase for women has been higher. Evidence of the negative impact of incarceration on the health of women of color suggests strategies to reduce these adverse effects. Correctional policies contribute to disparities in health between White women and women of color, providing a public health rationale for policy change. Specific roles for health professionals include becoming involved in alliances addressing alternatives to incarceration, creating programs that address the needs of women in correctional facilities, and identifying the pathways by which correctional policies damage health (Fredenberg, 2002).The study of Plugge, Douglas, and Fitzpatrick (2006) examined health of 505 women in prison through interview method. Significant difference was found between the female imprisoned time two and three and health service use. Therefore, the finding highlights the very poor health status of women prisoner as well as number of problems.

According to Science Daily (2011) the male are violent and angrier compared to their counterpart. Historically warriors have been made. Most interpersonal violent crimes have been done by men. Statistics reveal that 90% of murderers and 82% of other violent criminals are male. According to It consultant Suresh Shivaram, the anger is personal trait. Scratch the surface of some suave, sophisticated guys and you see seething, simmering discomfort and anger. Generally most of the men have a false sense of poor and ego. When they get angry, it is a sign of aggression. Even in victory as a sign of power. In even in wars we see victory as a sign of power. In society men are told to project their aggression and not hold back their anger. The masculine features, heavy brows and an angular face overlap with an angry expression. Analysis of corporate executive revealed that it is our own insecurities that drive us to aggression and violence. Psychologists D. Vaughan Becker believes that an angry man is one of the most dangerous characters around. A research by Harvard university reports revealed that women are less afflicted than men by overconfidence or the delusion that they know more than they really do. According to Anupam Kher "A man grows up thinking he is better off than the fairer sex. His parents flaunt it if he has many girlfriends. Men also have fewer outlets for their anger (Times News Network, 2011).

Intervention and Inmates Wellbeing

The psychological intervention can play pivotal role in maintaining the wellbeing of the people in general as well as jail imates in particular. Jin (1992) has observed the efficacy of Tai chi, a moving meditation, in reducing mood disturbance caused by mental/emotional stressors. Yoga is claimed to endow perfect physical, mental and social well being of an individual. A series of research investigations have revealed that there are many beneficial effects of yoga, which would help in the stress management (Selvamurthy, 1993). Yoga and meditation can contribute positively to various cognitive processes, including perception and in turn, on Subjective Well-being, Quality of Life and Criminal Propensity. Vipassana is a particular technique of self-examination, a scientific method of self-observation that results in the total purification of the mind and the highest happiness of full liberation (Vipassana Research Institute, 1990). This study examines gender similarities and differences in background characteristics, the effectiveness of treatment, and the predictors of post-release outcomes among incarcerated drug-using offenders. The sample of 1,842 male and 473 female treatment and comparison subjects came from a multi-site evaluation of prison-based substance abuse treatment programs. Three-year follow-up data for recidivism and post-release drug use were analyzed using survival analysis methods. Despite the greater number of life problems among women than men, women had lower three-year recidivism rates and rates of post-release drug use than did men. For both men and women, treated subjects had longer survival times than those who were not treated. There were both similarities and differences with respect to gender and the other predictors of the two post-release outcomes. Differences in background characteristics and in factors related to post-release outcomes for men and women suggest the plausibility of gender-specific paths in the recovery process. According to Aminabhai (1996) Yoga training leads to highly significant improvement in subject's mental health. Deepak, Manchanda, and Maheswari (1994) have reported that continuous meditation can substantially improve the clinico-electroencephalographic measures in drug resistant epileptics.Pal, Negi, Verma, Sapru, and Verma (1994) have observed that yoga practitioners, both males and females, have more positive attitude towards yoga than control group males and females. Control group showed higher neurotic trend, and yoga group showed significantly higher scores on social desirability. The authors have also found that Life event scores (past one year) were significantly less in yoga practitioners.

Jhansi, and Rao (1996) have investigated the role of practicing.

Transcendental Meditation (TM) in improving the attention regulation capacity of its practitioners. Their study reveals greater attention regulation capacity among TM practitioners compared to their counterparts, due to the regular cognitive exercises involved in meditation practice. Chandiramani, Verma, Dhar, and Aggarwal (1994) have studied the psychological effects of Vipassana meditation on Tihar jail inmates. They report that VM brought significant improvement in psychological parameters like sense of hope and well-being. There has been considerable reduction in the neurotic predisposition, hostility and feelings of helplessness reported by the prisoners. Mahendram, Kumariah, Mishra, and Baroohi (1998) have observed that VM is effective in reducing tension headache. Venkantesh.

U.S. Department of Justice (1995) studied effectiveness of treatment for drug abusers under criminal justice supervision. Rehabilitation and treatment offered to prison inmates proved a lot. Further therapeutic community model was found effective in preventing re-arrest.

According to Chandiramani et al (1995), Vipassana meditation emphasizes both conscious life style changes in the area of morality and deeper psychological analysis, which alters the contents and the processes of the mind in fundamental ways. Vipassana meditation courses have been found to bring out many positive changes in the behaviour of jail inmates (Shah, 1976; Unnithan & Ahuja, 1977; Hammersley & Creganj, 1986). On the basis of clinical experience Chandiramani et al (1995) have stated that mild to moderately severe neurotic cases of anxiety, depression and adjustment problems show complete recovery as a result of Vipassana. They also reported that there was considerable reduction in the neurotic predisposition, hostility and feelings of helplessness reported by the prisoners; while the sense of hope and well being were enhanced, following Vipassana courses. On the basis of the above review of research, it is assumed that Vipassana will have a significant positive effect on the Quality of Life, Subjective Well-being, and Criminal Propensity of undertrials.

As a coping mechanism, Koenig, (1995) reported that religion was the most important coping method while in prison for 32% of older men in prison. Inmates with strong belief in God were more likely to be serving their first prison term (56% vs. 29%), but were more likely to have multiple criminal charges (42% vs. 17%). Those inmates who engaged in daily prayer/Bible study were more likely to be a low security risk (90% vs. 74%), less likely to be on drug charges (39% vs. 77%), had shorter sentences (80 vs. 121 months). Prisoners with no religious affiliation and those whose parents had no religious affiliation had the highest depression scores.

A field experiment conducted by Khurana (1996, 1999) by using 'before and after' design to find out the effect of VM on Quality of Life (QOL) and Subjective Well-being (SWB) of undertrials, in Tihar jail. She found positive effect on VM on QOL and SWB of undertrials, though not significant. Therefore, she recommended that control group design should be used in further study. Chaudhary (1999) investigated the effectiveness of Vipassana meditation, as a technique of stress management and reformation among adolescent prisoners. In her study, Chaudhary reported that both state anxiety and trait anxiety reduced significantly among adolescents who had done the Vipassana course. She also reported that there was a decrease in aggression among undertrial prisoners who had undergone Vipassana course. There was an increase in the feelings of positive emotions such as, hopefulness, self control, conformity, and compassion, was more after practicing Vipassana, as compared to non-practitioners of Vipassana. In this article, the authors reflect on the lessons of their Stanford Prison Experiment, some 25 years after conducting it. They review the quarter century of change in criminal justice and correctional policies that has transpired since the Stanford Prison Experiment and then develop a series of reform-oriented proposals that still can be drawn from this and related studies on the power of social situations and institutional settings and can be applied to the current crisis in American corrections (Haney, & Zimbardo, 1998).

McArthur, Camilleri, and Webb. (1999) studied strategies for managing suicide and and self-harm in prisons. Since, there appear more self-harm in prison therefore psychological strategies in this regard in the form of counselling may prove better. Risk intervention team and High risk alert team may launch treatment package that in turn may prove effective for the same. Further prisoner support schemes and case management team may help a lot. Further there is a need to increase role of fellow prisoners and extension peer support. Recreational facilities be increased that include games, television, video, radio etc. Further up-gradation facilities to relieve overcrowding may prove useful. New system for monitoring punishment and adjudicating grievance, Prison routine, safety issues ad support services be provided to them. Prisoner handbook may be provided to them. Reforms and humanitarian approach be used and prisoner be given opportunity for basking light and safe shelter, appropriate blanket and cots be provided to them. Family support and visit be provided to them. Constructive activity within the prison be started. Support from prison staffs and probation officer be provided. Hopes and plans for future be provided. Inter-departmental communication be started. The staff that is professionally trained and valued be deputed.

The article of Klein and Bahr, (2004) evaluates a cognitive program at a state correctional institution. The program points to the benefits of cognitive skill development as inmates reflect on rejoining their families. Inmates are instructed on identifying problems, recognizing resources for solving problems, and thinking of alternative ways to respond as they apply principles of practical reasoning to everyday problems families experience. Mindfulness-based stress-reduction courses were offered in drug units in six Massachusetts Department of Corrections prisons. A total of 1,350 inmates completed the 113 courses. Evaluation assessments were held before and after each course, and highly significant pre- to post-course improvements were found on widely accepted self-report measures of hostility, self-esteem, and mood disturbance. Improvements for women were greater than those for men, and improvements were also greater for men in a minimum-security, pre-release facility than for those in four medium-security facilities. The results encourage further study and wider use of mindfulness-based stress reduction in correctional facilities.

The study of I. G. Surender Kumar organized a gathering of 2000 prisoner and allowed them to sing and dance that is perhaps the best efforts (Hafeez, 2011). The study mentioned here examines gender similarities and differences in background characteristics, the effectiveness of treatment, and the predictors of post-release outcomes among incarcerated drug-using offenders. The sample of 1,842 male and 473 female treatment and comparison subjects came from a multi-site evaluation of prison-based substance abuse treatment programs. Three-year follow-up data for recidivism and post-release drug use were analyzed using survival analysis methods. Despite the greater number of life problems among women than men, women had lower three-year recidivism rates and rates of post-release drug use than did men. For both men and women, treated subjects had longer survival times than those who were not treated. There were both similarities and differences with respect to gender and the other predictors of the two post-release outcomes. Differences in background characteristics and in factors related to post-release outcomes for men and women suggest the plausibility of gender-specific paths in the recovery process **(Pelissier, Camp, Gaes, Saylor & Rhodes, 2003).**

(Pelissier, Camp, Gaes, Saylor & Rhodes, 2003). Helping others definitely have positive impact on self. It was studied was conducted by Lebel (2008) on 228 incarcerated persons in prison reintegration programmes. The result revealed that the helper/wound healer's orientation has positive relation with self-esteem, and greater satisfaction with life and a negative relationship with having a criminal attitude and the forecast of the rarest. This study has brought together two seemingly socially extreme population subgroups to compare their health and social well-being.

These groups had in common restricted living arrangements and aspirational enrollment. As well, they are part of the population-based Thai Cohort Study (TCS) of 87,134 adult Open University students residing throughout the country. Analysis was restricted to men aged 20-39 years resulting in 711 monks, 195 prisoners and 29,713 other cohort members. For physical health, certain conditions such as tuberculosis or malaria much more common among prisoners, while goiter and liver diseases were more common among monks. This could be due to prison living arrangements for the former and region of residence for the latter. For other social outcomes, lower trust, higher economic stress and lower personal well-being was noted for prisoners compared to other groups. Findings here with regard to spirituality and religion are encouraging with almost no difference reported between prisoners and other cohort members implying that trust-building and other social intervention for prisoners could be activated through prevalent religious beliefs and practices and with continuing support from Thai prison authorities (Yiengprugsawan , Seubsman , & Sleigh, 2010).

The study of Esere, (2011) investigated the effectiveness of Six Thinking Hats Training in improving the psychological well- being of prisoners. The study consisted of 24 convicted prisoners whose ages ranged from 19-60 years with a mean of 22.50 years. A 2x2 experimental group design was adopted, and subjects were randomly assigned to the treatment and control groups. A validated instrument titled "Psychological Well-being Scale" was administered to the two groups before and after the experimental programme. The programme was packaged into 8 weekly workshop sessions. Analysis of Covariance was used to test the hypothesis at 0.05 level of significance. Results showed that the treatment package

(Six Thinking Hats Training) was effective in improving the psychological well-being of the prisoners, their age notwithstanding. Based on this finding, it was recommended that prisoners should always seek counsleling to enable them face the daunting challenges of incarceration and that Counsellors in Nigeria should diversify their services to non-school settings like the prison. So it can be concluded that the jail inmates experience more psychological vulnerabilities. Further the conditions of the inmates who already suffer from metal tensions is much worse than to their counterpart. Further it seems that the married people are more attached to their home therefore experience much mental distress compared to other but literature in this respect is less. Ii can be conducted just psychological interaction may more useful for promoting wellbeing among the jail inmates.

Hypotheses

1. The jail inmates will experience more depression as compared to their non-jail inmates' counterpart.
2. The married jail inmates will experience more depression as compared to their unmarried counterpart.
3. The jail inmates will report high anger as compared to the non-inmates counterpart.
4. The married jail inmates will report more anger as compared to their unmarried counterpart.
5. The jail inmates will be less satisfied from their life as compared to their counterpart.
6. The married jail inmates will be less satisfied from their life as compared to their counterpart.

7. The jail inmates will be poor in general health as compared to their non-jail inmate's counterpart.
8. The married jail inmates will be poor in general health as compared to their unmarried counterparts.
9. The jail inmates will be more religious as compared to their non-jail inmates counterparts.
10. The married jail inmates will be more religious as compared to their unmarried counterpart.
11. The jail inmates will be more involved in alcohol, smoking, and substances as compared to their non-jail inmates.
12. The married people will be more involved in alcohol, smoking, and substances compared to their unmarried counterpart.

ALCOHOL, SMOKING AND SUBSTANCE INVOLVEMENT SCREENING TEST

The objectives of the present study is to appraise psychological vulnerabilities among the jail inmates and non-inmates of Shimla. Further the study intends to assess and compare the psychological wellbeing among married and unmarried people in general as well as married jail inmates and non-inmates of this town. The intention behind their exploration is that the married people seems to be more attached to their families, wife and the children compared to the unmarried counterpart. Beside this the study has also tried to explore the qualitative information about the jail inmates who have incarcerated for the short period of time, moderate period of time and long period of time. Further the study also intends to see the relationship between birth order among the brother and its relation with criminality and wellbeing. But the analysis performed in the present study dealt only with married and unmarried jail inmates and non-inmates . The description of detailed methodology is as follows:-

Study Area

Presented study was conducted at Kanda jail witch is located in District Shimla which once was a capital of British people. The area selected for the present study was Kanda jail that is located in Ghanati which is approximately 20 Km from local bus stand Shimla. This Kanda Jail is appropriately 1.5 Km from Ghanati.

Sample

A sample consist of 120 (60 Jail Inmates + 60 Non-Inmates) subjects was selected and divided into two groups based on married that comprises of N = 30 in married and another N = 30 in unmarried categories. In this manner there were total four groups with N = 30 in each comprises of above mentioned total sample. The age of the subjects ranges from 30 to 60 years. The jail inmates were selected from Kanda Jail whereas the non-inmates were selected from Kanda village and Ghanahati areas.

Sample Classification

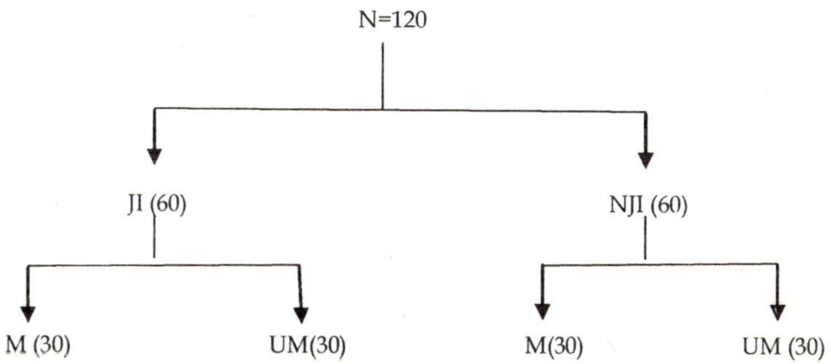

Notations: JI = Jail Inmates; NJI: Non-Jail Inmates; M = Married; UM: Unmarried.

Measures

In the study both qualitative and quantitative measures were used. The qualitative measures include naturalistic field observation, interview and photography whereas the qualitative measures include below mentioned self report inventories. The detailed descriptions of both the measures are as follows:

Field Observation

Field observation may be defined as making direct observation of a phenomenon under investigation by going to that very place in which it occurs. The research physically approached to the specific place with the help of local people as well as Municipal Corporation information and the areas being shown by National Sample Survey Organization and interacted with the people.

Interview

Interview is a face to face interaction between the interviewer and interviewee. It is a conversation, during which an interviewer seeks to elicit interaction about the subject to evaluate him in general terms. In an interview the interviewer has opportunity to observe certain aspects of subject's behaviour such as his manner of speaking his poise, and facial expressions etc. The researcher maintained face to face interaction with the people and maintained their record.

Case Study

A case study refers to an in-deep study of one person. In a case study, nearly every aspect of the subject's life and history is analyzed to seek patterns and causes for behavior. The hope is that learning gained from studying one case can be generalized to many others. It deals with collection and presentation of detailed information about a particular participant or small group, frequently including the accounts of subjects themselves. A form of qualitative descriptive research, the case study looks intensely at an individual or small participant pool, drawing conclusions only about that participant or group and only in that specific context.

Photography

Photography is the process of forming stable or permanent visible images either directly or indirectly by the action of light on sensitive surfaces. Photograph is used as a record of events in research work. In the present study the researcher formed certain images about the position of the squatter in the surroundings.

Short Depression-Happiness

It was developed by Stephan Joseph (2000). It is a 4 point scale. It has three positive items and three negative items is in this scale. Item number 1, 3, 6 has reverse scoring. The score ranges from minimum of 0 and maximum of 18.

The Anger Self-Report Questionnaire

It was developed by Nicola S Reynolds in 1994. The Anger Self-Report Questionnaire involves 30 items that involves 7 options that starts from strongly agree to strongly disagree.The score ranges from minimum of 30 to maximum 180.

Satisfaction with Life Scale

Diener, Emmons, Loven, Griffin, (1985). The SWLS is a short, 5 items instrument designed to measure global cognitive judgments on our lives. It involves 7 point scale that range from strongly disagree to strongly agree. More appropriately, it involves the scaling option strongly agree (1) disagree (2), slightly disagree (3), neither agree nor disagree (4), slightly agree (5) agree (6) and strongly agree

(7). The score range from minimum of 5 to a maximum of 35. Higher the score more satisfied the students will be in their day to day life.

PGI Health Questionnaire N-1

It was developed by Verma, Wig and Pershad in 1985. It has two sections. Section A consists of 16 questions and section B of 22 questions. The scores ranges from 0 to 16 in section, "A' and 0 to 22 in section 'B' with a total score of 38. Its reliability ranges from .86 to .88.

Religiosity Scale

This scale was developed by Dr. L. I. Bhusan, Professor of Bhagalpur University in 1990. It has 36 items. It is a five point scale where the score ranges from minimum of 36 to maximum of 180. Items No. 6, 8, 10, 13, 16, 19, 21, 27, 29, 34 and 35 have reverse scoring. The reliability of the scale ranges from .69 to .78.

Alcohol, Smoking and Substance Involvement screening test (ASSIST)

This test was developed by the 'World Health Organization' (WHO) in 2001. It contains 8 items those are related to different situations. It is a brief screening questionnaire used to find out people's involvement in alcohol, smoking and substances. Each question on the ASSIST has set of responses. Item No. one has only two option smoking and substances. Each question on the ASSIST has set of responses. Item one has only two options in the form of Yes and No whereas item No 2 to 6 has given option and remaining two items i.e. item No. 7 and 8 have four options. Thus the score ranges from 0 to 25. Higher the score more vulnerable the person will in the ASSIST. The test-re-test reliability coefficient ranges from .90 to .58.

Procedure

The objective of the present study was to appraise psychological vulnerabilities among the jail inmates and non-inmates of Shimla. Further the study intends to assess and compare the psychological wellbeing among married and unmarried people in general as well as married jail inmates and non-inmates in particular. The intention behind their exploration is that the married people are more attached to their families, wife and the children compared to the unmarried counterpart. Beside this the study has also tries to explore the qualitative information about the jail inmates who have incarcerated for the short period of time, moderate period of time and long period of time. Further the study also intends to see the relationship between birth order among the brother and its relation with criminality and wellbeing. But the analysis performed in the present study dealt only with married and unmarried jail inmates and non-inmates. For fulfilling the objective, a sample consist of 120 (60 Jail Inmates + 60 Non-Inmates) subjects was selected and divided into two comparable groups based on marriage that comprises of N = 30 in married and another N = 30 in unmarried categories. In this manner there were total four groups with N = 30 in each comprises of above mentioned total sample. The age of the subjects ranges from 30 to 60 years. The jail inmates were selected from Kanda Jail whereas the non-inmates were selected from Kanda village and Ghanahati areas. The Kanda Jail is further 1.5 Km from Ghanati. The researcher took the admission from the DIG and interviewed the aforesaid prisoner who were compared with the normal population of that areas i.e. from Ghanahati and Kanda village. These subjects were assessed with the help of qualitative tools that include observation, Interview, fuzzy cognitive mapping and photography techniques. Besides this psychological measures were also being used on them that include Short Depression-Happiness Scale. The Anger Self-Report Questionnaire, Satisfaction with life scale, PGI Health Questionnaire N-1. Religiosity Scale as well as Alcohal, Smoking and Substance Involvement screening test. A 2 X 2 way ANOVA has been used besides calculating mean and standard deviation. The score has been tabulated and displayed inappropriate table as shown in the next chapter.

Socio-Cultural Problem

In this chapter an attempt has been made to present the finding based on qualitative and quantitative observation. In the first section attempt has been made to present qualitative information whose description is as follows.

5.1 Qualitative Analysis

The qualitative analysis has been performed by using field observation, interview, case study and fuzzy cognitive mapping etc. During analysis the jail inmates reported physical, social, cultural as well as psychological problem whose description is as follows:

5.1.1 Physical Problem

The prison inmates of Kanda jail have to face various physical problems .The process inmates experience as they adjust to the lack of privacy, loss of freedom and rigid structure in prison. The entire prison experience with its symbolic mechanisms of justice that encompass every lock, piece of barbed wire, the thick walls, the never ending supervision and segregation, the harsh solitude, and minimalist lifestyles, are deliberately designed to not only incapacitate. The prison environment for them is terrific in Kanda jail. The prisoner here suffers from health problem. Food problem is the major problem. They only receive two time food. Nutritious food is absent. They have no recreational facilities. Some of them reported that they are not properly checked by the Medical expert during distress and disease. Their disease and suffering are always undermined. During sickness they do not receive appropriate treatment and social support from the government staffs as well as from their colleagues. They also suffer from dirt, dust and noise problem. The mosquito problem is also there and very few have the provision of net to escape them. Sometime the prison inmates are physically tortured by the police officer as well sometime there are infighting among them. Caste based discrimination is covertly and somewhere overtly practiced.

They got dinner at 5 P. M. which is very early and are deprived of night meal i.e. dinner 9 P.M. that exert its influence on their wellbeing. During minor illness they are not sent to the hospital for checking. Prison inmates lack choices about meals, free time to toll or who will be their cellmate. They also do not provide sufficient blankets. They have to lay down on cement bed that is not good. Further they are not hygienic and they do not receive appropriate hot water.

They receive very few water and not allowed to cook their own and insufficient fruits, food and vegetables in side the jail. They got tired in whole day and are not allowed to roam outside. Many prisoners witness violence and brutality and others become victimized. Fears about safety issues combined with constant loud noises can create much anxiety for prisoners. Fear of victimization causes many prisoners to become hyper vigilant. Difficult sleeping conditions can compound feelings of anxiety. Confinement can lead to feelings of panic and claustrophobia. For prisoners with a history of childhood trauma, incarceration can cause re-dramatization. The substance abuse problem is also there Some prisoner also are addicted to various substances. The drugs continue to be readily available in most prisons may be in Kanda jail.

There are problems of the safekeeping for all inmates in Kanda Jail. Safe keeping generally comprises of keeping inmates locked away, counted, and controlled whilst allowing for isolated moments of welfare activities to satisfy needs through recreation, education and counseling. Unfortunately, the welfare and psychological freedom of the individual inmate does not depend on how much education, recreation, and counseling he receives but rather, on how he manages to live and relate with the others inmates who constitute his crucial and only meaningful world. Therefore the

Fig 5.1: A view of Kanda jail

adjustment problem is the biggest problem. However in prison, inmates find themselves reduced to a level of living near bare subsistence.

The entire prison structure is based on solitude and separatism. Firstly, the convict is isolated from the external world and everything that motivated his/her offences. Secondly, they are to a large degree isolated from one another. Further the set patterns of clothing exert influence on them. The outbreaks of various diseases also affect them.

Timetables also play a large factor in rehabilitation by establishing rhythms, and cycles of repetition. This combined with convict's personal needs for reward and acquisition through penal labour, turns the criminal into a somewhat docile worker imposes on the convict the moral form of wages as the condition of his existence. A principle of order and regularity further worsened their life. The younger inmates are initially more resistant to the prison structure which makes them more likely to be the targets of victimization in comparison to older inmates who assume passive avoidance roles in prison .The police sometime use harsh word and excessive cruel mode such as punishment. However, it has been suggested that after the initial shock of imprisonment, younger inmates tend to demonstrate increasing levels of conformity over time. The need of personal fulfillment, privacy, chances of being attacked by other prisoners, less chance of self-advancement and self-improvement. The issue of freedom can be explored by minimizing restrictions and controlling prisoners at the same time. The inmate loses interest in the outside world, views the prison as home, loses the ability to make independent decisions, and in general, defines him or herself totally within the institutional context. There is a problem of power of roles, rules, symbols, group identity and situational validation of behaviour. They had a chain on one foot to remind them of the repression of being in prison. They experienced tired an physical breakdown; and sometime smashed their heads on the walls and refused to eat. The healthy males became too powerful and dominate other. They have to sit without work as a result become fat and come under the disease. Most of the prison inmates are fat, and have no work to do, no field to execute work. They have bear fluctuation weather of the Shimla.

5.1.2 Socio-cultural Problem

Prisoner has to face various social and cultural problems. They are not allowed to interact with the society, family and relatives. The society frame various rule and regulation to them. Further, the job problem is the major problem. After entering into the jail their character is blemished and spoiled and government and non-government organization hesitate to depute them on job. Marriage problem is another problem. No girl is ready to do marriage with the criminal. Further overage becomes another problem by which they are deprived of. The criminal record of the person also got influence on their family, spouse and children. The neighborer used to comment on them. It causes suppression among the children by which they are not able to explain their problem. The inmates are not involved in social and religious activities. Further they are seen with evil eyes. They face adjustment problem in the society. Some of them recall their past life and become compulsive to visit their nearer and dearer. They have to seek permission to go outside.

Prisoners are required to ask for permission to perform even some of the basic functions, such as asking to go to the toilet. Punishments that are given include solitary confinement, a diet of bread and water, as well as withdrawal of privileges such as cigarettes and recreation. The prisons do more harm than good. The pains of jail confinement affect all prisoners in different ways. To begin with the prisoners need to withstand the entry shock by adapting quickly to prison life. Prisoners are exposed to a new culture, which is very different from their own culture. Then they need to maintain outside links. For example, keeping in contact with family and friends becomes frustrating. For some prisoners the major source of stress would include the loss of contact with family and friends outside the prison. There is also the fear of deterioration. There is lack of personal choice within the prison environment which many effect prisoners. After many years of being told what to do they may well lose the ability to think for themselves and make their own decisions and choices freely

As prisoners experience barriers to contact with friends and family, many completely withdraw socially due to frustrations. Inmates also experience decreased feelings of personal value as they no longer engage in meaningful daily activities. Lack of purpose, combined with lack of leisure time activities and the absence of positive social interactions, causes feelings of depression and despair for many prisoners. The culture of prisons discourages inmates from showing many emotions. Expressions of sadness can be viewed as a weakness and vulnerability. Toughness, anger and hostility tend to be considered an asset. Many inmates join gangs to avoid being exploited. Other inmates engage in violent crime to gain acceptance. It has also been recognized, through simulations of prison environments, that lockups and isolation have the habit of dehumanize prisoners by making them feel anonymous, and breeding ill feelings because of their rejection and condemnation by society as a whole. Likewise, it must be remembered that offenders have been drawn from a society in which possessions are closely linked with concepts of personal worth by numerous cultural definitions. They are not allowed to interact with society. They have fear that society will seem them evil eyes. They are quite distressed from the law and order of the society and wrong judgment of the court as well as are quite worried about the money and muscle power in the society that inhibit the truth and even honest person have to go inside the jail and wicked one comes out. They are also worried about the sycophancy and link of powerful people to the leader as well as growing corruption as well as social discrimination.

5.1.3 Psychological Problem

The imprisonment can be brutal, demeaning, and generally psychologically a devastating experience for many prison inmates. Psychological symptoms may include psychosis, severe depression, inhibiting anxiety, and complete social withdrawal. Another major stressor which the prisoner is faced with in prison is the fear of contagious and incurable diseases, such as, aids. The prison is extremely stressful, at the other extreme, those who are dependent, passive, and generally incompetent may find that the prison structure offers them a positive experience. Prisons are often the scenes of brutality, violence and stress. Prisoners are faced with incidence of violence and are always concerned for their safety. It is not an emotional or psychological fear.

Many prison inmates report a variety of psychological changes as a result of the stress associated with incarceration. The effects of prison can create long-lasting psychological changes that cause difficulties for some inmates even after they have been released. Other inmates develop post traumatic stress disorder as a result of the stressful conditions. Loss of autonomy about almost all aspects of daily activity can decrease feelings of self-worth. The inmates who do not receive psychological treatment remain at a high risk of repeating criminal offenses after release. This process becomes a survival skill for many inmates but it prevents them from being able to experience and express true emotions. Sometime poor habit and trait of their colleagues caused problem to them. The psychological significance as to the prisoner's conception of his personal adequacy. There are also some mentally deferent who are maltreated .

The use of solitary confinement has adverse psychological effects among the prisoner such as: delusions, dissatisfaction with life, claustrophobia, depression, feelings of panic, and on many instances madness. Further symptom of chronophobia – a state often referred to as prison neurosis is also visible. Prisons issued uniforms also play a large part in destroying personal identity, and crashing individual spirits. The inmates are put in isolation. These include disciplinary segregation, administrative segregation, and protective custody. As one may predict, whether an inmate becomes severely psychologically effected by solitary confinement is dependent on how much time is spent in isolation.These somewhat bland, yet repetitive outfits are a way whereby unification maybe achieved within inmates, through the portrayal that they are no longer individuals, but are part of a whole. That whole is symbolic of - society. A situational withdrawal, an intransigent line where prisoners refuse to cooperate with the staff and show hostility towards the institution and solitary confinement is some other issues.

The psychological deprivations such as loss of liberty; moral rejection; restriction to obey rules and sexual deprivation places pressure on prisoners towards homosexual satisfaction of one's sexual needs. It also involves a cry for the compassion of a woman. Also loss of autonomy suggests that prisoners are under the control of officials. Prisoners must obey rules and are treated like children. Combined these psychological deprivations lead to a destruction of the human personality.

The emotional disruption and adjustment are the problems for most inmates during the early stages of their sentences, resulting from the dramatic disruptions to their life caused by the many restrictions, deprivations and constraints inherent in prisons. Stress levels may increase at the end of an inmate's sentence due to anticipation and feelings of uncertainty about one's ability to adjust and cope in the outside world again, after having adjusted to prison life. After conviction they have also psychological fear that other people will see them as wicked. Some of the jail inmates practice caste system inside the jail and therefore dominate other and quarrel with each other.

The prisoner of Kanda jail recalled about their past life, prefer loneliness and do not share their feeling with other that leads to suppression and psychological and stressful problem. They always have evil thoughts about the crime they did. They are not satisfied with their life and the activities they have done. The memories of their family, spouse, and children career as well as socio-economic difficulties of their own and family further make them worrisome. They make un-necessary plan to get rid of prison and want to do any work by their thought processes that further make them in problem. Some of them become neurotic and psychotic due to their psychological problem.

Some of them consider themselves as very superior and expect other to follow them. They have hallucination and illusion. Some of them reported superiority complex whereas other reported as inferiority complex. Some of them are very gossiping and try to befool them.

5.2 Quantitative Analysis

5.2.1 Analysis of Happiness and Depression among Jail Inmates

Happiness is commonly defined as a stat of well being that is commonly associated with feeling good or experiencing pleasure . Understanding happiness and related positive mental states involves the study of contentment with the past, happiness in the present and hoop for the future. There is not one key that can lead to happiness and life satisfaction; instead, multiple components are involved in finding happiness. This is similar to having different ingredients in a recipe for happiness. The social relationship can greatly influence your happiness. People who have close relationship with family and friend tend to have grater life satisfaction and happiness. More appropriately, the happiness is not just the absence of depression, but also the presence of a number of positive emotional and cognitive states. But the depression deals with severe despondency and dejection, accompanied by feelings of hopelessness and inadequacy. A condition of mental disturbance, typically with lack of energy and difficulty in maintaining concentration or interest in life. The happiness depression analysis have been shown in table 5.1.

From the table 5.1 it is quite clear that the main effect of Institution on the measure of Short Depression and happiness measure was found $F(1, 116) = 18.80$, $p<.001$ as highly statistically significant. The non-jail inmates experienced more depression whereas the jail inmates experience less depression therefore; hypothesis No. 1 which states that "*The jail inmates will experience more depression as compared to their non-nail inmate's counterpart*" is rejected. It can be viewed from the F value as well as shown in fig 5.1.

Table 5.1 A 2 × 2 way ANOVA performed on Married and Unmarried Jail Inmates and Non-Inmates on Short Depression Happiness Scale

Source	SS	df	*ms*	F	*p*
Total	30173.000	120			
I	126.075	1	126.075	18.80	p<.001
M	3.675	1	3.675.548	.548	n.s.
I × M	31.008	1	31.008	4.62	p<.05
Error	777.833	116	6.705		

Notations: I = Institution; M= Marriage.

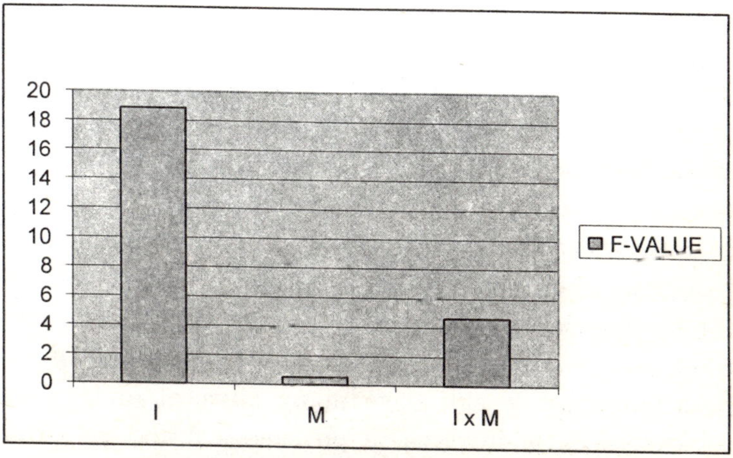

Fig. 5.2 Showing F values among married and unmarried jail inmates and non-inmates on depression and happiness parameter

The main effect of Marriage was found F (1, 116) = .548, p>.05 as statistically non-significant. But the interaction between I X M was found F (1, 116) = 4.62, p<.05 as statistically significant. A non significant difference was found between the married and unmarried jail inmates in the parameter of depression. Therefore, hypothesis No. 2 "The married jail inmates will experience more depression as compared to their unmarried counterpart" is rejected.

5.2.2 Appraisal of Anger among Jail Inmates

Anger is an emotion related to one's psychological interpretation of having been offended, wronged or denied and a tendency to undo that by retaliation. It is a normal emotion that involves a strong uncomfortable and emotional response to a perceived provocation. That has three modalities: cognitive (appraisals), somatic-affective (tension and agitations) and behavioral (withdrawal and antagonism).It is like pressure cooker; we can only apply pressure against our anger for a certain amount of time until it explodes. It may have physical correlates as well such as increased heart rate, blood pressure, and levels of adrenaline and nor adrenaline. Some view anger as part of the fight or flight brain response to the perceived threat of harm. It becomes the predominant feeling behaviorally, cognitively, and physiologically when a person makes the conscious choice to take action to immediately stop the threatening behavior of another outside force. The English term originally comes from the term *anger* of Old Norse language. It can have many physical and mental consequences. It involves a physiological and emotional response we have when we believe a cherished idea, or concept has been violated. The biological parameters of anger can include; A change in heart rate, breath rate, hormones, muscle tension and neurotransmitter etc. In anger there are is a physiological and emotional response we have when we believe a cherished idea, or concept has been violated Therefore it is an automatic response to ill treatment. The analysis of married and unmarried jail inmates and non inmates is shown on table 5.2.

Table 5.2 A 2 X 2 way ANOVA performed on Married and Unmarried Jail Inmates and Non-Inmates on Anger Self-Report

Source	SS	df	*ms*	F	*p*
Total	1719236.000	120			
I	821.633	1	821.633	5.71	p<..01
M	9.633	1	9.633	.067	n.s.
I × M	5.633	1	5.633	.039	n.s.
Error	16.698.267	120			

Notations: I = Institution; M = Marriage

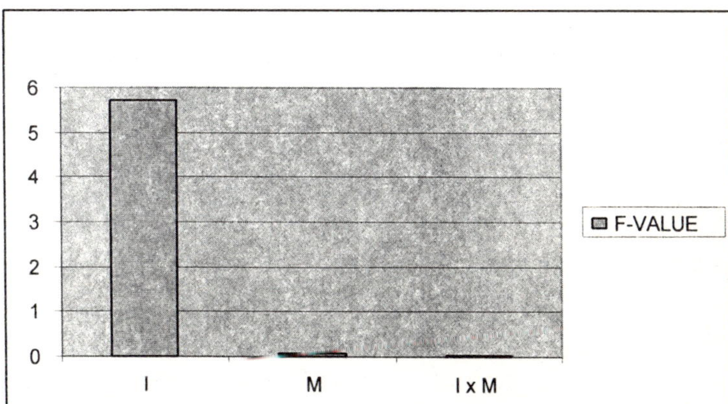

Fig. 5.3 Showing F values among married and unmarried jail inmates and non-inmates on Anger

From the table 5.2 it is quite clear that the main effect of Institution on the measure of anger self-report scale F(1, 116) = 5.71, p<.01 as statistically significant. The jail inmates reported more anger as compared to their counterpart. Therefore hypothesis No. 3 which states that "*The jail inmates will be high in anger as compared to the non-inmates group*" is accepted. The main effect of Marriage was found $F(1, 116) = .067$, $p>.05$ as statistically non-significant.

The interaction between I × M was found $F(1, 116) = .03$, $p>.05$ as statistically non-significant. There is very few difference among the groups in the parameter of anger therefore hypothesis No. 4 which states that "The married jail inmates will report more anger as compared to their unmarried counterpart" is rejected. Although there slight difference between all the group and ANOVA has showed non-significance yet there exists slight difference and married jail inmates have reported more anger compared to their counterpart. Further the analysis of satisfaction with there life is as follows.

5.2.3 Analysis of Life Satisfaction among Jail Inmates

Life satisfaction refers to how people evaluate their life as a whole rather than their current feelings. It captures a reflective assessment of which life circumstances and conditions are improved for subjective wellbeing. It ranges from positive to negative affect. The satisfaction is a Latin word that means to make or do enough. Satisfaction with one's life implies contentment with or acceptance of one's life circumstances, or the fulfillment of one's wants and needs for one's life as a whole. In essence, life satisfaction is a subjective assessment of the quality of one's life. Because it is inherently an evaluation, judgments of life satisfaction have a large cognitive component. According to Ed Diener and his colleagues (1999), subjective well-being,or happiness, has both an affective and a cognitive (i.e.,4judgmental) component. The affective component consists of how frequently an individual reports experiencing positive and negative affect. It is considered to be the cognitive component of this broader construct. Researchers differentiate between life-domain satisfaction and life-as-a whole (or global) life satisfaction. Life-domain dealt with with specific areas of an individual's life, such as work, marriage, and income, whereas judgments of global life satisfaction are much more broad, consisting of an individual's comprehensive judgment of her life. The analysis is as follows :

Table 5.3 A 2 × 2 way ANOVA performed on Married and Unmarried Jail Inmates and Non-Inmates on Satisfaction with Life Scale

Source	SS	df	*ms*	F	*p*
Total	42803.000	120			
I	2193.075	1	2193.075	154.326	p<.001
M	18.408	1	18.408	1.295	n.s.
I x M	27.075	1	27.075	1.905	n.s.
Error	1648.433	116	14.211		

Notations: I = Institution; M = Marriage (1)

From the table 5.3 it is quitc clear that the main effect of Institution on the measure of Satisfaction with life was found F(1, 116) = 154.33, p<.001 as statistically significant. The jail inmates showed less satisfaction from their life therefore hypothesis No. 5 which states that "*The jail inmates will be less satisfied from their life compared to their non-jail inmates*" is accepted.

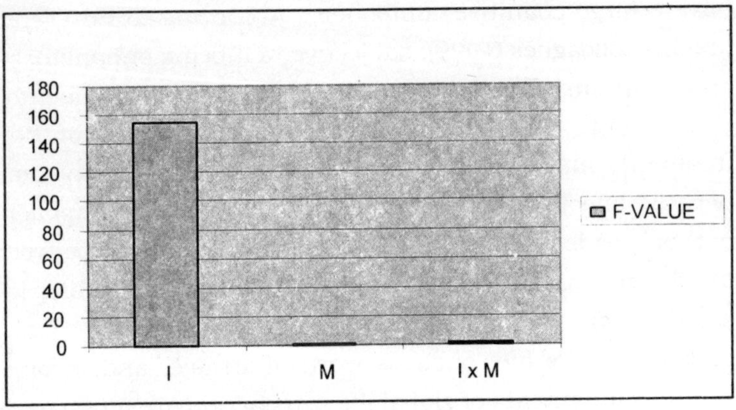

Fig 5.4: Showing F values among married and unmarried jail inmates and non-inmates on life satisfaction

The main effect of Marriage was found F (1, 116) = 1.295, p>.05 as statistically non-significant. But the interaction between I X M was found F (1, 116) = 1.905, p>.05 as statistically significant. Married jail inmates have lowest score in the life satisfaction as compared to other group therefore hypothesis No. 6 which states that *"The married jail inmates will be less satisfied from their life as compared to their unmarried counterpart"* is rejected. Although the interaction effects are non-significant yet the score of married jail inmates is very less compared to other groups.

5.2.4 : Appraisal of Health

Health is a state of complete physical, mental and social wellbeing of a person and not merely the absence of disease or infirmity. Mental health describes either a level of cognitive or emotional well-being or an absence of a mental disorder. From perspectives of the discipline of positive psychology or holism mental health may include an individual's ability to enjoy life and procure a balance between life activities and efforts to achieve psychological resilience. It is an expression of emotions and signifies a successful adaptation to a range of demands. The World Health Organization defines mental health as "a state of well-being in which the individual realizes his or her own abilities, can cope with the normal stresses of life, can work productively and fruitfully, and is able to make a contribution to his or her community". It was previously stated that there was no one "official" definition of mental health. Cultural differences, subjective assessments, and competing professional theories all affect how "mental health" is defined. There are different types of mental health problems, some of which are common, such as depression and anxiety disorders, and some not so common, such as schizophrenia and Bipolar disorder. The analysis performed on general health is as follows :-

Table 5.4 A 2 × 2 way ANOVA performed on Married and Unmarried Jail Inmates and Non-Inmates of PGI-health

Source	SS	df	*ms*	F	*p*
Total	27699.000	120			
I	1.875	1	1.875	.158	n.s.
M	5.208	1	5.208	.440	n.s.
I x M	3.675	1	3.675	.310	n.s.
Error	1373.833	116	11.843		

Notations: I = Institution; M = Marriage

From the table 5.4 it is quite clear that the main effect of Institution on the measure of PGI-health questionnaire Area (a) was found F (1, 116) = .158, p>.05 as statistically significant. There is very less difference between the two groups, therefore hypothesis No. 7 which states that *"The jail inmates will be poor in general health as compared to their non-jail inmate's counterpart"* is rejected. The main effect of Marriage was found F(1, 116) = .440, p>.05 as statistically non-significant.

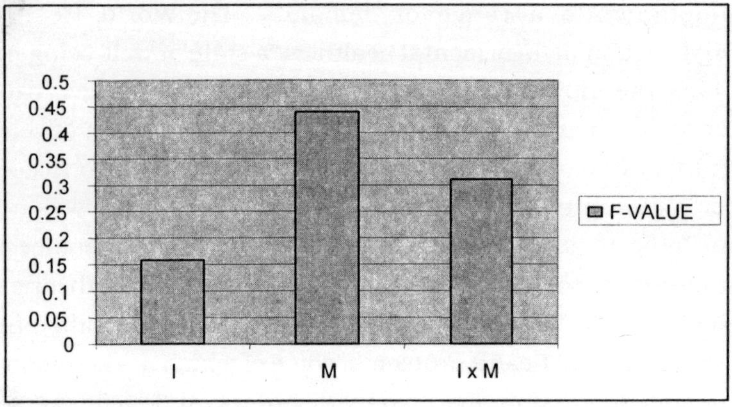

Fig. 5.5 Showing F values among married and unmarried jail inmates and non-inmates on general health

But the interaction between I X M was found F(1, 116) = .310, p>.05 as statistically significant. There is very less difference among the entire group. Therefore, hypothesis No. 8 which states that *"The married jail inmates will be poor in general health as compared to their counterpart unmarried counterparts"* is rejected. The analysis on health Area-A is as follows :

Table 5.5 A 2 × 2 way ANOVA performed on Married and Unmarried Jail Inmates and Non-Inmates on PGI-health questionnaire Area -A

Source	SS	df	*ms*	F	*p*
Total	7287.000	120			
I	3.675	1	3.675	.697	n.s.
M	8.008	1	8.008	1.519	n.s.
I x M	18.408	1	18.408	3.492	p<.05
Error	611.500	116	5.272		

Notations: I = Institution; M = Marriage

From the table 5.5 it is quite clear that the main effect of Institution on the measure of Wellbeing-Area-A was found

F (1, 116) = .697, p>.05 as statistically significant. The main effect of Marriage was found F (1, 116) = 1.519, p> .05, as statistically non-significant.

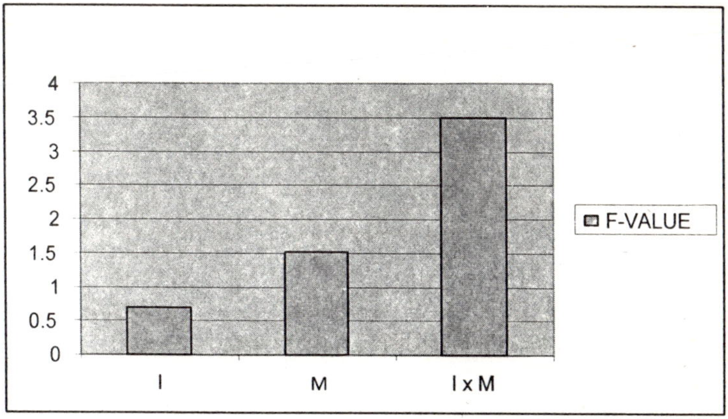

Fig. 5.6: Showing F values among married and unmarried jail inmates and non-inmates on general health (Area-A)

But the interaction between I × M was found F (1, 116) = 3.492, p< .05 statistically significant. There is very little difference between all the groups. Yet interaction is significant.

Table 5.6 A 2 × 2 way ANOVA performed on Married and Unmarried Jail Inmates and Non-Inmates on PGI-health questionnaire Area-B

Source	SS	df	*ms*	F	*p*
Total	7159.000	120			
I	10.208	1	10.208	1.924	n.s.
M	.408	1	.408	.077	n.s.
I x M	6.075	1	6.075	1.145	n.s.
Error	615.433	116	5.305		

Notations: I = Institution; M = Marriage

From the table 5.6 it is quite clear that the main effect of Institution on the measure of was found F(1, 116) = 1.92, p> .05 as statistically significant.. The main effect of Marriage was found F(1, 116) = .077, p>.05 as statistically non-significant.

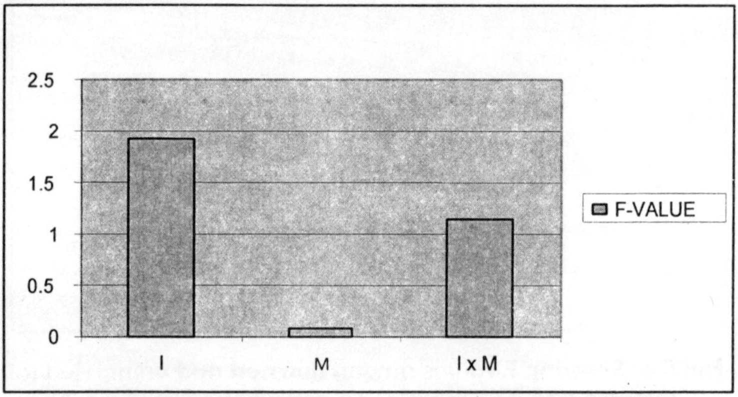

Fig. 5.7 Showing F values among married and unmarried jail inmates and non-inmates on general health (Area-B)

But the interaction between I × M was found F (1, 116) = 1.145, p>.05 as statistically significant. There is very little difference between all the groups therefore hypothesis No. 7 and 8 are rejected.

5.2.4 Analyses Religiosity

The **religiosity** refers to the *possession by a person or group of some inner spiritual piety* that may or may not lead on to attachment to a religious practice or a pre-defined set of beliefs but have **some outward social impact. It** refers to the numerous aspects of religious activity, dedication, and belief (religious doctrine). In its narrowest sense, religiosity deals more with how religious a person is, and less with how a person is religious (in practicing certain rituals, retelling certain myths, revering certain symbols, or accepting certain doctrines about deities and afterlife). Similarly, the spirituality refers to experience of a deep seated sense of meaning and purpose in life together with a sense of belonging. It is about acceptance, integration and wholeness. The spiritual wellbeing is an integral part of mental, emotional and physical health. It is considered to be a primary coping resource on the journey of recovery and healing. The spirituality and faith provide an opportunity to detach from circumstances and observe life with clarity and integrity. It can either be positive or negative. The analysis performed to know religiosity is as follows:

Table 5.7 A 2 X 2 way ANOVA performed on Married and Unmarried Jail Inmates and Non-Inmates on Religiosity Scale

Source	**SS**	**df**	***ms***	**F**	***p***
Total	1887496.000	120			
I	1020.833	1	1020.833	5.57	p<.01
M	163.333	1	163.333	.892	n.s.
I x M	61.633	1	61.633	.337	n.s.
Error	21236.867	116	183.076		

Notations: I = Institution; M = Marriage

From the table 5.7 it is quite clear that the main effect of Institution on the measure of spiritual wellbeing measure was found $F(1, 116) = 5.57$, $p<.01$ as statistically significant.. The jail inmates were found more religious compared to their counterpart. Therefore, hypothesis No. 9 which states that the "The jail inmates will be more religious as compared to their non-jail counterparts" is accepted. The main effect of Marriage was found $F(1, 116) = .892$, $p>.05$ as statistically non-significant.

0

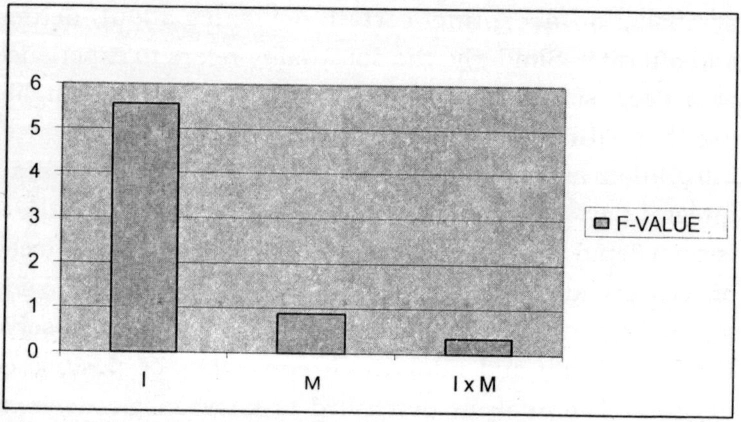

Fig. 5.8 Showing F values among married and unmarried jail inmates and non-inmates on Religiosity

The interaction between I × M was found $F(1, 116) = .337$, $p > .05$ as statistically non-significant. Unmarried jail inmates were found more religious therefore hypothesis No. 10 which states that "The married jail inmates will be more religious as compared to their unmarried counterpart" is rejected.

5.2.5 Analysis of Substance Abuse

Substance abuse can simply be defined as a pattern of harmful use of any substance for mood-altering purposes. Medline's medical encyclopedia defines drug abuse as "the use of illicit drugs or the abuse of prescription or over-the-counter drugs for purposes other than those for which they are indicated or in a manner or in quantities other than directed."But the broad range of substance abuse in today's society is not that simple. There are substances that can be abused for their mood-altering effects that are not drugs at all — inhalants and solvents — and there are drugs that can be abused that have no mood-altering or intoxication properties, such as anabolic steroids. Generally, when most people talk about substance abuse, they are referring to the use of illegal drugs. Most professionals in the field of drug abuse prevention argue that any use of illegal drugs is by definition abuse. Those drugs got to be illegal in the first place because they are potentially addictive or can cause severe negative health effects; therefore, any use of illegal substances is dangerous and abusive. Others argue that casual, recreational use of some drugs is not harmful and is merely use, not abuse. The most vocal of the proponents of recreational drug use are those who smoke marijuana. They argue that marijuana is not addictive and has many beneficial qualities, unlike the "harder" drugs.

Table 5.8 A 2 × 2 way ANOVA performed on Married and Unmarried Jail Inmates and Non-Inmates on Substance Abuse Scale

Source	SS	df	*ms*	F	*p*
Total	4673.000	120			
I	35.208	1	35.208	2.78	.05
M	.208	1	.208	.016	n.s.
I × M	11.408	1	11.408	.898	n.s.
Error	1474.300	116	12.709		
Total					

Notations: I = Institution; M = Marriage

From the table 5.8 it is quite clear that the main effect of Institution on the measure of substance abuse scale was found F(1, 116) = 2.78, p<.01 as statistically significant. The non inmates reported more involvement in substances as compared to the jail inmates therefore hypothesis No. 11 which states that "The jail inmates will be more involved in alcohol, smoking, and substances as compared to their non-jail inmates" is rejected.

The main effect of Marriage was found F (1, 116) = .016, p>.05 as statistically non-significant. It has been shown in fig 5.8

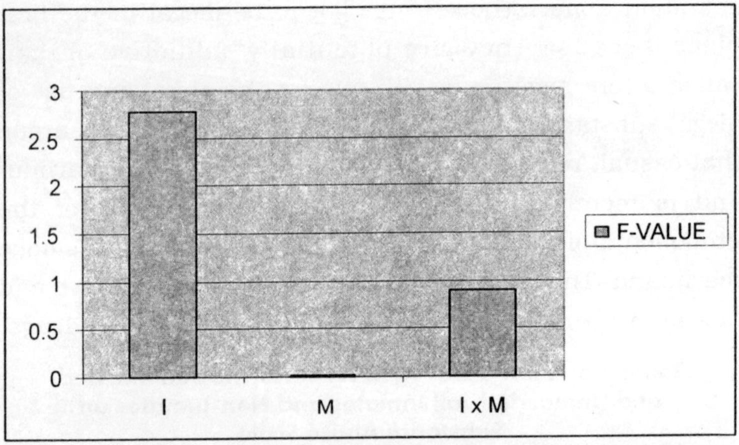

Fig. 5.9: Showing F values among married and unmarried jail inmates and non-inmates on substance abuse

The interaction between I × M was found F(1, 116) = .898, p>.05 as statistically non-significant. The married people reported less involvement in ASSIST compared to other group. Therefore hypothesis No. 12 which states that "The married people will be more involved in alcohol, smoking, and substances compared to their unmarried counterpart "is also rejected.

MARRIED AND UN-MARRIED JAIL INMATES

In this chapter an attempt has been made to compare jail inmates and non mates with married and unmarried marital status by computing their mean score. Objective of the present chapter is to compare jail inmates especially of married and unmarried on depression, anger, wellbeing, life satisfaction and substance abuses. Since, the jail inmates suffer from depression, suicidal, anxiety, morality, guilt and delinquent behaviour, therefore it is important to appraise their psychological vulnerabilities by analyzing their mean score on each psychological measure. In the first table an attempt has been made to compared married and unmarried jail inmates by comparing them to non-inmates on depression and happiness parameter.The average score have been shown in table 6.1, description is as follows:

Table 6.1 Average Score of Married and Un-Married Jail Inmates and Non-jail-Inmates on Depression Happiness Parameter

Groups	Married	Unmarried	Average
Jail Inmates	15.27	13.90	14.585
Non-Inmates	16.96	16.30	16.63
Average	16.115	15.1	15.60

From the Table No. 6.1 it is quite clear that the average score of the inmates on the short depression scale was found as 14.585 whereas the average score of non-inmates was founds as 16.63. It shows that the jail inmates have more level of depression as compared to their counterpart. Similarly the average score of married was 16.115 whereas the average score of unmarried was 15.1 It shows that unmarried have more anger as compare to their counterpart. The average score of married inmates on the anger self-report scale was found as 15.27 whereas the average score of unmarried inmates was founds as 13.90. It shows that the married inmates have more level of depression as compared to their counterpart. The average score of married non-inmates on the anger self-report scale was found as 16.96 whereas the average score of unmarried non-inmates was founds as 16.30. It shows that the married non-inmates have slightly more depression as compared to their counterpart. However the computed statistics has already showed the difference as statistically non-significant.

The average score of married inmates on the short depression happiness scale was found as 15.27 whereas the average score of unmarried non-inmates was founds as 16.96. It shows that the -married inmates have slightly more level of depression as compared to their counterpart. However the computed statistics has already showed the difference as statistically non-significant. The average score of married inmates on the short depression happiness scale was found as 13.90 whereas the average score of unmarried non-inmates was founds as 16.30. It shows that the -Unmarried non-inmates have slightly more level of depression as compared to their counterpart. However the computed statistics has already showed the difference as statistically non-significant.

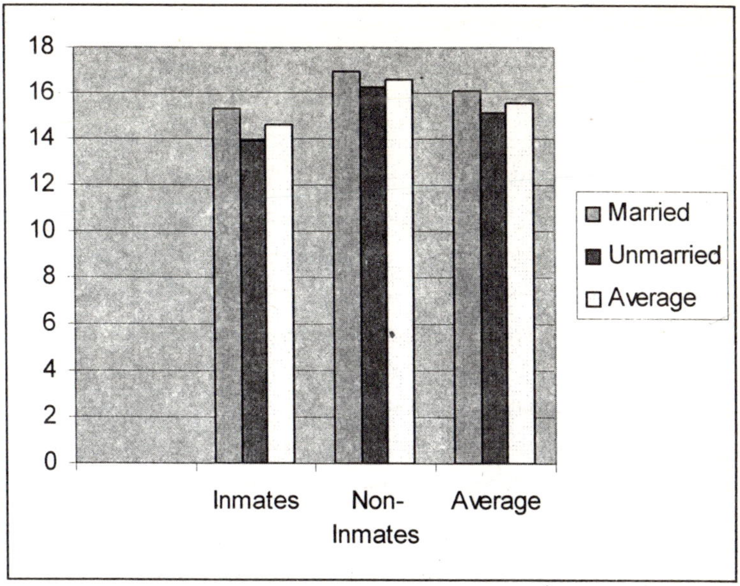

Table 6.1 Profile of married and un-married jail inmates and non-jail-inmates on the depression -happiness parameter

Average score of married inmates was found 15.27 whereas the average score of unmarried non-inmates was found 16.30 It shows that married inmates have more depression as compared to their counterpart. Average score of married non-inmates was found 16.96 whereas the average score of unmarried inmates was found 13.90. It shows that the unmarried inmates have more depression as compared to their counterpart.

Suicidal death and criminal activities in the prison have increased in past one decades (McArthur, M., Camilleri, & Webb, 1999). According to the study of Ivanoff, Smyth, Grochowski, Jang, and Klein (1992) the jail inmtes suffer from means–ends problem solving and suicidality. It is most prevalent among adult male prison inmates. Krishna (1993) reports the presence of neuroticism, anxiety, extraversion, and morality guilt among adolescents who are high on delinquent behaviour. According to Ogloff, Roech, & Hart (1994), united state has the largest per capita prison population in the industrialized world. Here 6.5% to 10% inmates suffer from serious mental illness and 15% to 40% suffer from moderate mental illnesses. There are adverse psychological consequences of overcrowding (Haney, 2001) in the prison. According to **James. D. J., & Glaze, L.(2006) the mental health problem has increased. Its example are** American jail inmates where Unprecedented number of are suffering from mental disorder and waiting for appropriate services have increased (Steadman, H. J. & Veysey, B. M. (1997). According to Suter, J. & Byrne, M. (2000 the female offender are different from male offender.

Table 6.2 Average Score of Married and Un-Married Jail Inmates and Non-jail-Inmates on the Anger Self-Report

Groups	Married	Unmarried	Average
Jail Inmates	121.20	122.20	121.7
Non-Inmates	116.53	116.40	116.46
Average	118.86	119.3	119.08

From the Table No. 6.2 it is quite clear that the average score of inmates on the anger self-report scale was found as 121.7 whereas the average score of non-inmates was founds as 116.46. It shows that the jail inmates have more level of anger as compared to their counterpart. Similarly the average score of married was 118.86 whereas the average score of unmarried was 119.3. It shows that unmarried have more anger as compare to their counterpart. The average score of married inmates on the anger self-report scale was found as 121.20 whereas the average score of unmarried inmates was founds as 122.20. It shows that the unmarried jail inmates have more level of anger as compared to their counterpart. The average score of married non-inmates on the anger self-report scale was found as 116.53 whereas the average score of unmarried non-inmates was founds as 116.40. It shows that the married non-inmates have slightly more level of anger as compared to their counterpart. However the computed statistics has already showed the difference as statistically non-significant.

The average score of married inmates on the anger self-report scale was found as 121.20 whereas the average score of unmarried Married non-inmates was founds as 116.53. It shows that the –married jail inmates have slightly more level of anger as compared to their counterpart. However the computed statistics has already showed the difference as statistically non-significant. 6. The average score of married inmates on the anger self-report scale was found

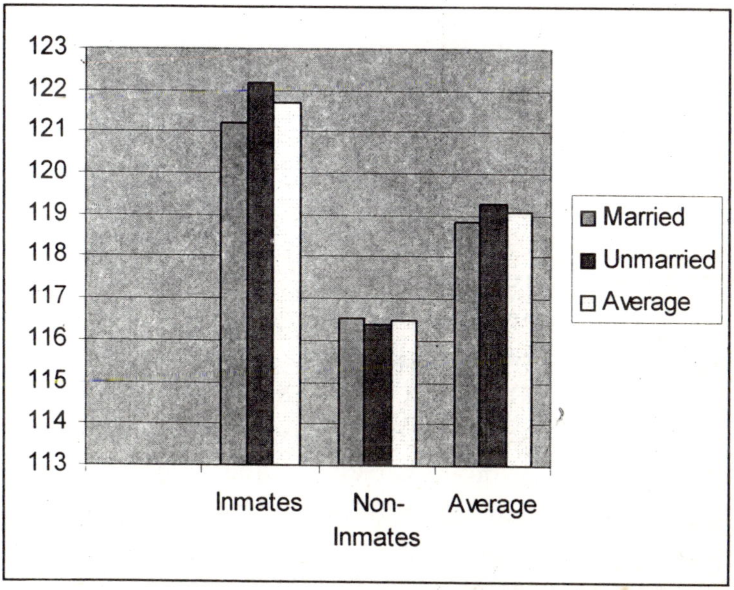

Table 6.2 Average score of married and un-married jail inmates and Non-jail-inmates on the anger self report

as 122.20 whereas the average score of unmarried non-inmates was founds as 116.40. It shows that the – Unmarried non-inmates have slightly more level of anger as compared to their counterpart. However the computed statistics has already showed the difference as statistically non-significant.

Average score of married jail inmates was found 121.20 whereas the average score of unmarried non-inmates was found 116.40. It shows that married jail inmates have more anger as compared to their counterpart. The married inmates perhaps are more attached to there family. Average score of married non-inmates was found 116.53 whereas the average score of unmarried inmates was found 122.20. It shows that the unmarried inmates have more anger as compared to their counterpart. The unmarried people are more worried about there life.

The mental disorder and violent offending is the serious problem in the jail (Mitchell (1999). The prisoner suffers from physical, social, and emotional health care needs, including specific needs related to their expressions of sexuality while in prison (Dumond, 2000). Behavioral or emotional disorder associated with paternal incarceration (Gabel, 1994). The study of Osofsky (1996) reveals the presence of certain psychological or personality factors. Further antisocial personality characteristics **(Jon & Gisli, 1996);** low self-esteem as an important cause of violence (Baumeister, Smart, L. & Boden, 1996). Similarly psychological distress as the main reason **(Pamela, & Frank, 2000) behind criminal activities. According to Howells, Day, Bubner, Jauncey, Williamson, Parker and Heseltine (2002), the criminal suffer from anger that become very difficult to control.** According to Smith, Parish, (2010) When a person with mental illness goes to prison they have to experience more worse treatment compared to other.

Table 6.3 Average Score of Married and Un-Married Jail Inmates and Non-jail-Inmates on the Satisfaction with Life

Groups	Married	Unmarried	Average
Jail Inmates	12.87	14.60	13.735
Non-Inmates	22.20	22.36	22.28
Average	17.535	18.48	18.00

From the Table No. 6.3 it is quite clear that the average score of inmates on the Satisfaction with life scale was found as 13.735 whereas the average score of non-inmates was founds as 22.28. It shows that the non inmates have more satisfied as compared to their counterpart. The problem of their personal life may be important areas. Similarly the average score of married was 17.535 whereas the average score of un-married was 18.48. It shows that Unmarried have more anger as compare to their counterpart. The average score of married jail inmates on the Satisfaction with life scale was found as 12.87 whereas the average score of unmarried inmates was founds as 14.60. It shows that the unmarried inmates have more level of anger as compared to their counterpart .The unmarried jail inmates are more worried about jail life.

The average score of married non-inmates on the Satisfaction with life scale was found as 22.20 whereas the average score of unmarried non-inmates was founds as 22.36. It shows that the unmarried non-inmates have slightly more level of anger as compared to their counterpart. However the computed statistics has already showed the difference as statistically non-significant. The average score of married inmates on the Satisfaction with life scale was found as 12.87 whereas the average score of married non-inmates was founds as 22.20. It shows that the -married non-inmates have slightly more good level of satisfaction as compared to their counterpart. However the computed statistics has already showed the difference as statistically non-significant. The average score of married inmates on the Satisfaction with life scale found as 14.60 whereas the average score of unmarried non-inmates was founds as 22.36. It shows that the – unmarried non-inmates have slightly more level of satisfaction as compared to their counterpart. However the computed statistics has already showed the difference as statistically non-significant

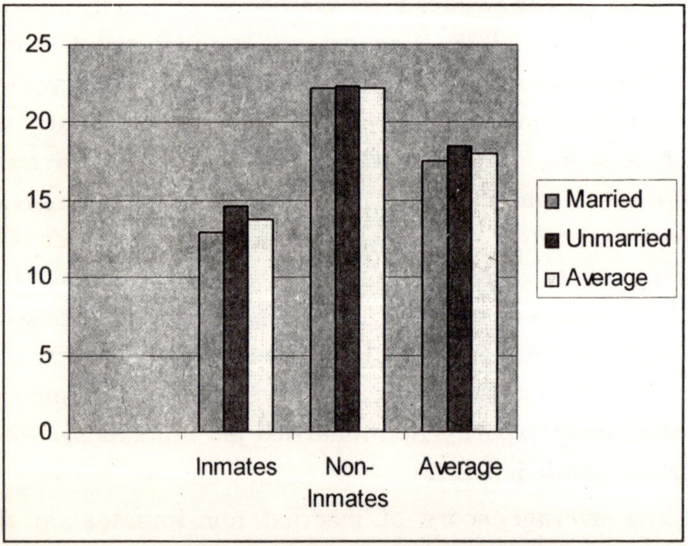

Table 6.3 Profile of married and un-married jail inmates and non-jail-inmates on the satisfaction with life

Average score of married inmates was found 12.87 whereas the average score of unmarried non-inmates was found 22.36. It shows that unmarried non-inmates have more anger as compared to their counterpart. Average score of married non-inmates was found 22.20 whereas the average score of unmarried inmates was found 14.60. It shows that the married non-inmates were more satisfied as compared to their counterpart .The prisons are often the scenes of brutality, violence and stress (Tosh, 1982:86) and are less satisfied from their life. Aside from problems with vision and hearing, their physical health did not appear to be any worse when compared to that of younger inmates (Elaine & Gallagher, 1990). The prisoner or jail inmates are prone to various social stressors **(Pelissier, Camp, Gaes, Saylor & Rhodes, 2003).** The jail inmates sometime die inside the jail that put question mark on the facilities being provided by the authorities inside the prison (Kamal, 2010). Several inmates after inquiring reported that medical facilities here are not significant (Sharma, 2011). They suffer from prison sexual violence in correctional facilities (Tess & Carl, 2010). Some inmates died during clinical trials (Sinha, 2011). According to Sen (2011) good inmates are those who get open air in jail. A cloud of dust and smoke hanging over the area, air and ater pollution and non functional sewerage plants are responsible for it (Thomas, 2011). The high levels of drug involvement, but considerable variation in severity of use and health and social consequences((Belenko, & Peugh, 2004).

Table 6.4 Average Score of Married and Un-Married Jail Inmates and Non-jail-Inmates on PGI Health Questionnaire (Total)

Groups	Married	Unmarried	Average
Jail Inmates	14.97	14.90	14.935
Non-Inmates	14.30	15.06	14.68
Average	14.635	14.98	14.8

From the Table No. 6.4 it is quite clear that the average score of inmates on the PGI Health Questionnaire N-1 was found as 14.935 whereas the average score of non-inmates was founds as 14.68. It shows that the non inmates were healthier as compared to their counterpart. Similarly the average score of married was 14.635 whereas the average score of unmarried was 14.98. It shows that married were healthier as compare to their counterpart. The average score of married jail inmates on the PGI Health Questionnaire N-1 was found as 14.97 whereas the average score of unmarried jail inmates was founds as 14.90 . It shows that the married inmates have healthier as compared to their counterpart. The average score of married non-inmates on the PGI Health Questionnaire N-1 was found as 14.30 whereas the average score of unmarried non-inmates was founds as 15.06. It shows that the married non-inmates were slightly healthy as compared to their counterpart. The average score of married inmates on the PGI Health Questionnaire N-1 was found as 14.97 whereas the average score of unmarried non-inmates was founds as 14.30.

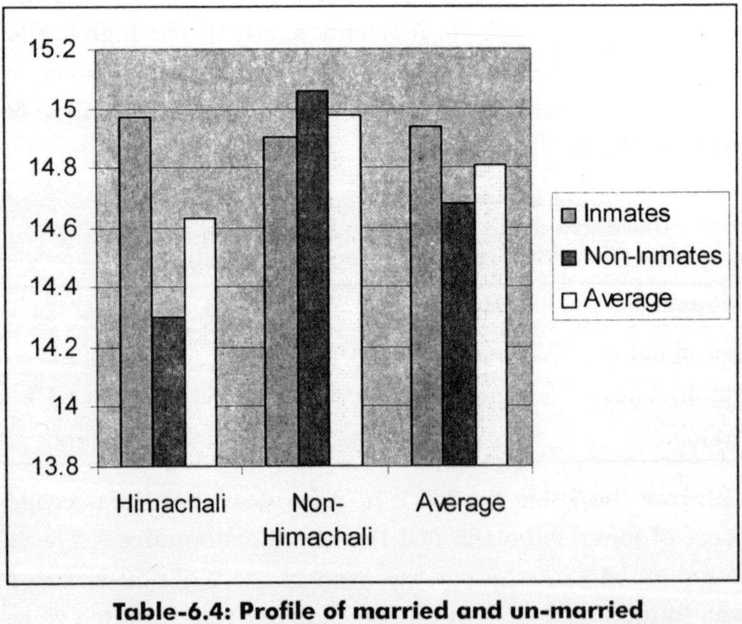

Table-6.4: Profile of married and un-married jail Inmates and non-jail-inmates on PGI Health

It shows that the -married noninmates were slightly healthy as compared to their counterpart. The average score of married inmates on the PGI Health Questionnaire N-1 was found as 14.90 whereas the average score of unmarried non-inmates was founds as 15.06. It shows inmates were slightly more healthier as compared to their counterpart. However the computed statistics has already showed the difference as statistically non-significant.

Average score of married jail inmates was found 14.97 whereas the average score of unmarried non-inmates was found 15.06. It shows that married jail inmates have better health as compared to their counterpart. Average score of married non-inmates was found 14.30 whereas the average score of unmarried inmates was found 14.90. It shows that the unmarried inmates have more good health as compared to their counterpart. **The wellbeing issues is major concern among the prisoner. According to Morgan, Steffan, Shaw, and Wilson, (2007) the prison inmates suffer from** mental health services, physical health concerns, negative affect, interpersonal relationships, and institutional relations) and four types of potential barriers to service utilization (self-preservation concerns, procedural concerns, self-reliance, and professional service provider concerns).

Table 6.5 Average Score of Married and Un-Married Jail Inmates and Non-jail-Inmates on PGI Health Questionnaire (Area-A)

Groups	Married	Unmarried	Average
Jail Inmates	7.10	7.40	7.25
Non-Inmates	6.96	8.20	7.58
Average	7.03	7.8	7.415

From the Table No. 6.5 it is quite clear that the average score of inmates on the PGI Health Questionnaire N-1 was found as 7.25 whereas the average score of non-inmates was founds as 7.58. It shows that the jail inmates have better health as compared to their counterpart. Similarly the average score of married people was 7.03 whereas the average score of unmarried was 7.8. It shows that unmarred people are healthy as compare to their counterpart. Similarly the average score of married inmates on the PGI Health Questionnaire was found as 7.10 whereas the average score of unmarried inmates was founds as 7.40. It shows that the unmarried inmates have slightly better health as compared to their counterpart. Further the average score of married non-inmates on the PGI Health Questionnaire N-1 was found as 6.96 whereas the average score of unmarried non-inmates was founds as 8.20. It shows that the married non-inmates are slightly healthy as compared to their counterpart. The average score of married inmates

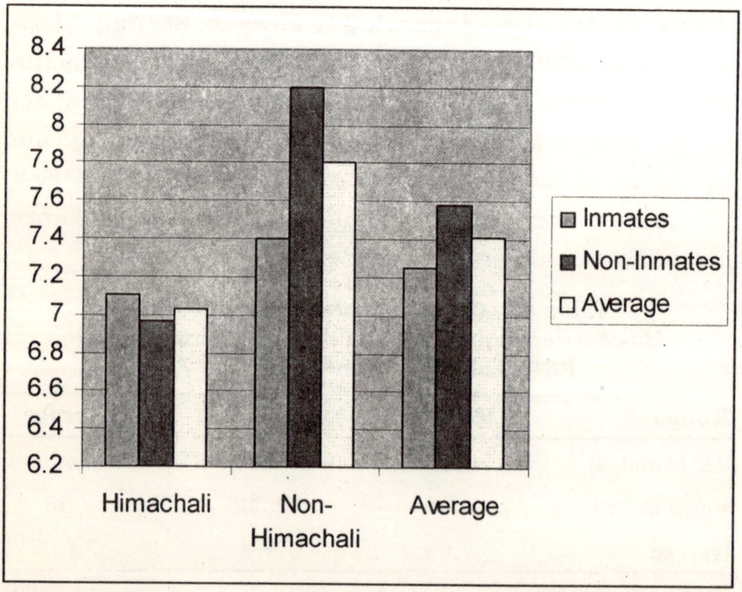

Table 6.5 Profile of married and un-married jail inmates and non-jail-inmates on PGI Health Questionnaire (Area-A)

on the PGI Health Questionnaire N-1 was found as 7.10 whereas the average score of married non-inmates was founds as 6.96. It shows that the –married jail inmates were slightly more good in healthy as compared to their counterpart. However the computed statistics has already showed the difference as statistically non-significant.

The average score of married inmates on the PGI Health Questionnaire N-1 was found as 7.40 whereas the average score of unmarried non-inmates was founds as 8.20. It shows that the –unmarried non-inmates were slightly more healthy as compared to their counterpart. Similarly the average score of married inmates was found 7.10 whereas the average score of unmarried non-inmates was found 8.20. It shows that married inmates have better health as compared to their counterpart. Further average score of married non-inmates was found 6.96 whereas the average scores of unmarried inmates was found 7.40. It shows that the unmarried inmates were healthier as compared to their counterpart. The prisoner suffers from various bodily disease and health problem. Gender-based violence - including rape, domestic violence, mutilation, murder, and sexual abuse further worsen their health (World Bank, 1993). Growing prison populations in the U.S. are largely due to drug-related crime and drug abuse (Belenko, & Peugh, 2004). The study reveals that the jail inmates suffer from health compared to counter part.

Table 6.6 Average Score of Married and Un-Married Jail Inmates and Non-jail-Inmates on PGI Health Questionnaire (Area-B)

Groups	Married	Unmarried	Average
Inmates	7.83	7.50	7.665
Non-Inmates	7.37	6.80	7.085
Average	7.6	7.15	7.375

From the Table No. 6.6 it is quite clear that the average score of inmates on PGI Health Questionnaire N-1 was found as 7.66 whereas the average score of non-inmates was founds as 7.08. It shows that the inmates were healthier as compared to their counterpart. Similarly the average score of married was 7.6 whereas the average score of unmarried people was 7.15. It shows that un-married people were healthier as compare to their counterpart. The average score of married inmates on the anger self-report scale was found as 7.83 whereas the average score of unmarried inmates was founds as 7.50. It shows that the unmarried inmates were more healthy as compared to their counterpart. Further the average score of married non-inmates on PGI Health Questionnaire N-1 was found as 7.37 whereas the average score of unmarried non-inmates was founds as 6.80. It shows that the married non-inmates were slightly healthy as compared to their counterpart. The average score

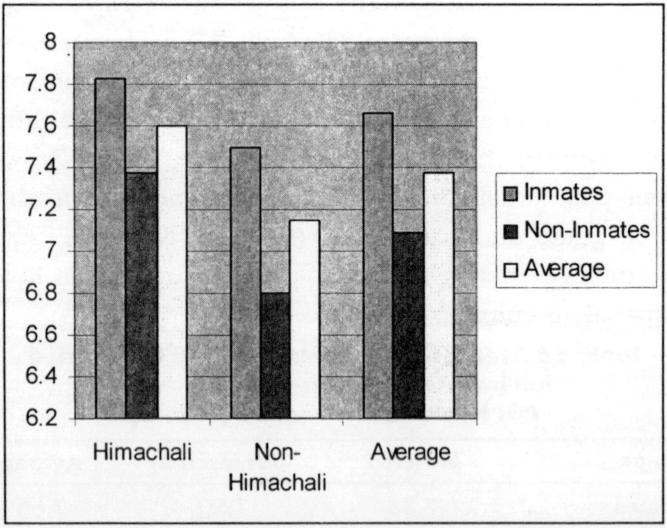

Fig. 6.5 Profile of married and un-married jail inmates and non-jail-inmates on PGI Health Questionnaire (Area-B)

of married jail inmates on PGI Health Questionnaire N-1 was found as 7.83 whereas the average score of married non- inmates was founds as 7.36. It shows that the -married inmates were slightly more healthy as compared to their counterpart. Similarly the average score of married inmates on PGI Health Questionnaire N-1 was found as 7.83 whereas the average score of unmarried non-inmates was founds as 7.50. It shows that the married jail inmates were slightly healthier as compared to their counterpart. However the computed statistics has already showed the difference as statistically non-significant.

Average score of married inmates was found 7.83 whereas the average score of unmarried non-inmates was found 6.80. It shows that married inmates were more healthy as compared to their counterpart. Average score of married non-inmates was found 7.37 whereas the average score of unmarried inmates was found 7.50. It shows that the unmarried jail inmates were healthier as compared to their counterpart. There are various studies witch shows that the jail inmates suffer from health and mental health problems. For example 64% of jail inmates (Doris & Lauren, 2006) suffering from mental health problem. There are certain factors e.g. more social conformity and close friends, and less need for employment counseling, felony drug convictions, drug dealing income, and unprotected sex than dropouts that is associated with crime. Completers were also less likely to encounter recent problems with significant other, have a psychiatric history, experience gunshot or stabbing, and commenced heroin use at older ages than dropouts. However, completers reported higher alcohol use than counterparts. The "life choice" (substance use, criminal and sexual behavior), static (background and dispositional), and dynamic situational influences **(Michelle & Steven, 2000) further aggravate their problem.**

Table 6.7 Average Score of Married and Un-Married Jail Inmates and Non-Jail-Inmates on Religiosity

Groups	Married	Unmarried	Average
Jail Inmates	127.13	128.03	127.58
Non-Inmates	123.63	119.86	121.745
Average	125.315	123.945	124.66

From the Table No. 6.7 it is quite clear that the average score of inmates on Religiosity was found as 127.58 whereas the average score of non-inmates was founds as 121.745. It shows that the inmates have more level of religious as compared to their counterpart. Similarly the average score of married was 123.315 whereas the average score of unmarried was 123.945. It shows that Unmarried have more religious as compare to their counterpart. The average score of married inmates on Religiosity scale was found as 127.13 whereas the average score of unmarried inmates was founds as 128.03. It shows that the unmarried inmates have more level of religious as compared to their counterpart.

The average score of married non-inmates on Religiosity scale was found as 123.63 whereas the average score of unmarried non-inmates was founds as 119.86. It shows that the married non-inmates have slightly more religious as compared to their counterpart. However the computed statistics has already showed the difference as statistically non-significant. The average score of married inmates on Religiosity scale was found as 127.13 whereas the average score of married non-inmates was founds as 123.63. It shows that the -married inmates have slightly more religious as compare to their counterpart. However the computed statistics has already showed the difference as statistically non-significant.

The average score of married inmates on Religiosity scale was found as 128.03 whereas the average score of unmarried non-inmates was founds as 119.86. It shows that the – married inmates have slightly more religious as compared to their counterpart.

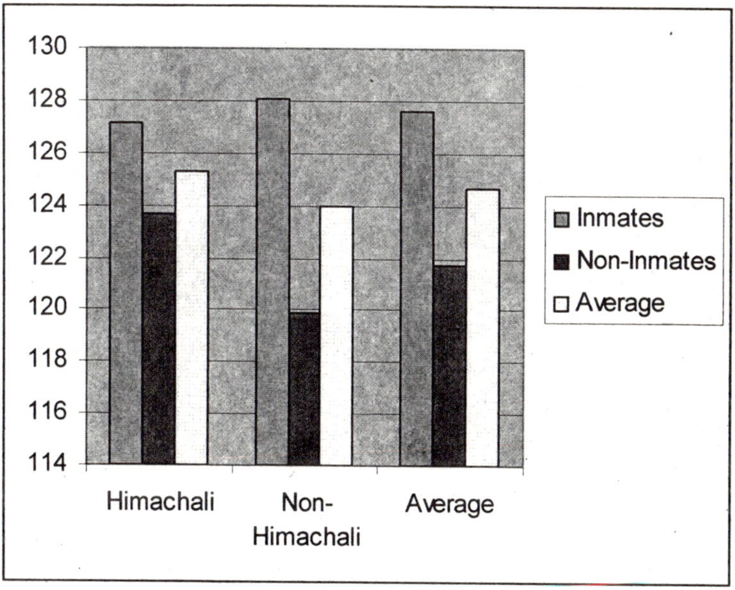

Fig 6.7 Average score of married and un-married jail inmates and non-jail-inmates on religiosity

Average score of married inmates was found 127.13 whereas the average score of unmarried non-inmates was found 119.86. It shows that married inmates have more anger as compared to their counterpart. Average score of married non-inmates was found 123.63 whereas the average score of unmarried inmates was found 128.03. It shows that the unmarried inmates have more religious as compared to their counterpart.

Table 6.8 Average Score of Married and Un-Married Jail Inmates and Non-jail-Inmates Alcohol, Smoking and Substance Involvement

Groups	Married	Unmarried	Average
Inmates	4.93	4.23	4.58
Non-Inmates	5.93	5.40	5.66
Average	5.43	4.815	5.1225

From the Table No. 6.8 it is quite clear that the average score of inmates on Alcohol, Smoking and Substance Involvement Screening test was found as 4.58 whereas the average score of non-inmates was founds as 5.66. It shows that the jail inmates have more level of substance as compared to their counterpart. Similarly the average score of married was 5.43 whereas the average score of unmarried was 4.81. It shows that unmarried have more substance as compare to their counterpart. The average score of married inmates on Alcohol, Smoking and Substance Involvement was found as 4.93 whereas the average score of unmarried inmates was founds as 4.23. It shows that the unmarried inmates have more level of substance as compared to their counterpart.

The average score of married non-inmates on Alcohol, Smoking and Substance Involvement was found as 5.93 whereas the average score of unmarried non-inmates was founds as 5.40. It shows that the married non-inmates have slightly more level of substance as compared to their counterpart. However the computed statistics has already showed the difference as statistically non-significant. The average score of married inmates Alcohol, Smoking and Substance Involvement was found as 4.93 whereas the average score of unmarried Himachali non-inmates was founds as 5.93. It shows that the – married non-inmates have slightly more level of substance as compared to their counterpart. The average score of married inmates on Alcohol, Smoking and Substance Involvement Screening test was found as 4.23 whereas the average score of unmarried non-inmates was founds as 5.40. It shows that the – unmarried non-inmates have slightly more level of substance involvement as compared to their counterpart.

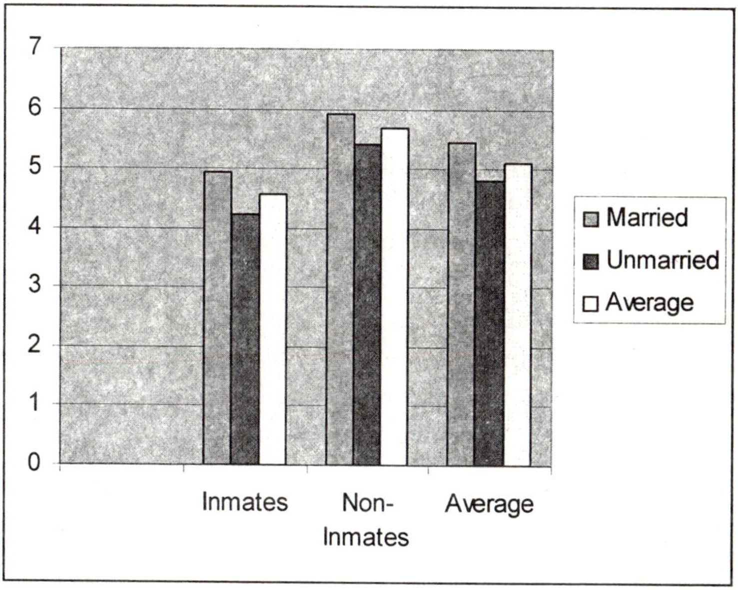

Table 6.8 Profile of married and un-married jail inmates and non-jail-inmates on substance involvement

Average score of married inmates was found 4.93 whereas the average score of unmarried non-inmates was found 5.40. It shows that married inmates have more substance as compared to their counterpart. Average score of married non-inmates was found 5.93 whereas the average score of unmarried inmates was found 4.23. It shows that the unmarried inmates have more substance as compared to their counterpart. **The proportion of substance abuser in the criminal justice systems are very high (Travis, J. & Visher, C. (1995) among the criminal. It can be associated with self-harm (Delly, (2006).** According to Sinha (2011) by 2020, 15% of world ailments will be due to mental disorders. The substances and drug related problem is the major reason. It may lead according to Baumeister, Catanese, and Wallace (2002) unwanted sexual activity. There are sexual assault in prisons (Wolff, Shi, Blitz, & Siegel, 2007).

According to Elaine and Gallagher (1990) insecurity as the major problem. For avoiding the samel the prisoner may use substances. There are negative impact of incarceration on the health(Freudenberg, 2002) for which the inmates use alcohol. The rehabilitation (Van Marle, 2007) problem is also associated with their poor mental health problem. The inmates have higher suicide rates than their community (Zhang, Grabiner, Virginia, Zhou & Li, 2010). These socio-psychological problem may be overcome through Yoga training (Aminabhai, 1996); continuous meditation (Deepak, Manchanda, and Maheswari, 1994); Transcendental Meditation (Jhansi, and Rao (1996). The role of practicing Vipassana meditation (Chandiramani, Verma, Dhar, and Aggarwal, 1994; Mahendram, Kumariah, Mishra, and Baroohi, 1998). Further, the study of Venkantesh, Pal, Negi, Verma, Sapru, and Verma (1994) have observed that yoga practitioners may be effective for reducing psychological vulnerabilities. The study of Koenig, (1995) focues on belief in God. That may useful for promoting well-beig. McArthur, M., Camilleri, P. & Webb, H. (1999) studied strategies for managing suicide and and self-harm in prisons. *The article of Klein and Bahr, (2004) also evaluates a cognitive program for* stress reduction and correctional facilities (2007).I. G. Surender Kumar organized a gathering of 2000 prisoner and allowed them to sing and dance that is perhaps the best efforts (Hafeez, 2011). Such as innovative programs may prove better resources for ameliorating psychological vulnerabilities among the prisoner.

CONCLUSION

The objective of the present study was to appraise psychological vulnerabilities among the jail inmates and non-inmates of Shimla. Further the study intends to assess and compare the psychological wellbeing among married and unmarried people in general as well as married jail inmates and non-inmates in particular. The intention behind their exploration is that the married people are more attached to their families, wife and the children compared to the unmarried counterpart. The study has been conducted in Kanda Jail on a sample of 120 (60 Jail Inmates + 60 Non-Inmates) subjects was selected and divided into two groups based on married that comprises of N = 30 in married and another N = 30 in unmarried categories. In this manner there were total four groups with N = 30 in each comprises of above mentioned total sample. The age of the subjects ranges from 30 to 60 years. The jail inmates were selected from Kanda Jail whereas the non-inmates were selected from Kanda village and Ghanahati areas. Both qualitative and quantitative measures were used. The qualitative measures includes Field observation, Interview, Case study, Photography, Short Depression-Happiness scale, The Anger Self-Report Questionnaire, Satisfaction with Life Scale, PGI Health Questionnaire N-1, Religiosity Scale, and Alcohol,

Smoking and Substance Involvement screening test. A 2 X 2 way ANOVA has been used besides calculating mean and standard deviation. The score has been tabulated and displayed inappropriate table as shown in the next chapter.

A 2 × 2 ANOVA results revealed that the main effect of Institution on the measure of Short Depression and happiness measure was found $F(1, 116) = 18.80$, $p<.001$ as highly statistically significant. The non-jail inmates experienced more depression whereas the jail inmates experience less depression therefore; hypothesis No. 1 which states that *"The jail inmates will experience more depression as compared to their non-jail inmate's counterpart"* is rejected. The main effect of Marriage was found $F(1, 116) = .548$, $p>.05$ as statistically non-significant. But the interaction between I × M was found $F(1, 116) = 4.62$, $p<.05$ as statistically significant. A non significant difference was found between the married and unmarried jail inmates in the parameter of depression. Therefore, hypothesis No. 2 *"The married jail inmates will experience more depression as compared to their unmarried counterpart"* is rejected. More appropriately, the average score of the inmates on the short depression scale was found as 14.585 whereas the average score of non-inmates was founds as 16.63. It shows that the jail inmates have more level of depression as compared to their counterpart. Similarly the average score of married was 16.115 whereas the average score of unmarried was 15.1. It shows that unmarried have more anger as compare to their counterpart. The average score of married inmates on the anger self-report scale was found as 15.27 whereas the average score of unmarried inmates was founds as 13.90. It shows that the married inmates have more level of depression as compared to their counterpart. The average score of married non-inmates on the anger self-report scale was found as 16.96 whereas the average score of unmarried non-inmates was founds as 16.30. It shows that the married non-inmates have slightly more depression as compared to their counterpart. However the computed statistics has already showed the difference as statistically non-significant. The average score of married inmates on the short depression happiness scale was found as 15.27 whereas the average score of unmarried non-inmates was founds as 16.96. It shows that the -married inmates have slightly more level of depression as compared to their counterpart. However the computed statistics has already showed the difference as statistically non-significant. The average score of married inmates on the short depression happiness scale was found as 13.90 whereas the average score of unmarried non-inmates was founds as 16.30 It shows that the – Unmarried non-inmates have slightly more level of depression as compared to their counterpart. However the computed statistics has already showed the difference as statistically non-significant. Average score of married inmates was found 15.27 whereas the average score of unmarried non-inmates was found 16.30 It shows that married inmates have more depression as compared to their counterpart. Average score of married non-inmates was found 16.96 whereas the average score of unmarried inmates was found 13.90. It shows that the unmarried inmates have more depression as compared to their counterpart.

As per the parameter of anger is concern the main effect of Institution on the measure of anger self-report scale $F(1, 116) = 5.71$, $p<.01$ as statistically significant. The jail inmates reported more anger as compared to their counterpart. Therefore hypothesis No. 3 which states that "*The jail inmates will be high in anger as compared to the non-inmates group*" is accepted. The main effect of Marriage was found F(1, 116) = .067, p>.05 as statistically non-significant. The interaction between I X M was found F (1, 116) = .03, p>.05 as statistically non-significant. There is very few difference among the groups in the parameter of anger therefore hypothesis No. 4 which states that "The married jail inmates will report more anger as compared to their unmarried counterpart" is rejected. Although there slight difference between all the group and ANOVA has showed non-significance yet there exists slight difference and married jail inmates have reported more anger compared to their counterpart.More appropriately, the average score of inmates on the anger self-report scale was found as 121.7 whereas the average score of non-inmates was founds as 116.46. It shows that the jail inmates have more level of anger as compared to their counterpart. Similarly the average score of married was 118.86 whereas the average score of unmarried was 119.3. It shows that unmarried have more anger as compare to their counterpart. The average score of married inmates on the anger self-report scale was found as 121.20 whereas the average score of unmarried inmates was founds as 122.20. It shows that the unmarried jail inmates have more level of anger as compared to their counterpart. The average score of married non-inmates on the anger self-report scale was found as 116.53 whereas the average score of unmarried non-inmates was founds as 116.40. It shows that the married non-inmates have slightly more level of anger as compared to their counterpart. However the computed statistics has already showed the difference as statistically non-significant. The average score of married inmates on the anger self-report scale was found as 121.20 whereas the average score of unmarried Married non-inmates was founds as 116.53. It shows that the –married jail inmates have slightly more level of anger as compared to their counterpart. However the computed statistics has already showed the difference as statistically non-significant. 6. The average score of married inmates on the anger self-report scale was found as 122.20 whereas the average score of unmarried non-inmates was founds as 116.40. It shows that the – Unmarried non-inmates have slightly more level of anger as compared to their counterpart. However the computed statistics has already showed the difference as statistically non-significant. Average score of married jail inmates was found 121.20 whereas the average score of unmarried non-inmates was found 116.40. It shows that married jail inmates have more anger as compared to their counterpart. The married inmates perhaps are more attached to there family. Average score of married non-inmates was found 116.53 whereas the average score of unmarried inmates was found 122.20. It shows that the unmarried inmates have more anger as compared to their counterpart. Unmarried are more worried about there life.

The main effect of Institution on the measure of Satisfaction with life was found $F(1, 116) = 154.33, p<.001$ as statistically significant. The jail inmates showed less satisfaction from their life therefore hypothesis No. 5 which states that "*The jail inmates will be less satisfied from their life compared to their non-jail inmates*" is accepted. The main effect of Marriage was found $F(1, 116) = 1.295, p>.05$ as statistically non-significant. But the interaction between I x M was found $F(1, 116) = 1.905, p>.05$ as statistically significant. Married jail inmates have lowest score in the life satisfaction as compared to other group therefore hypothesis No. 6 which states that "*The married jail inmates will be less satisfied from their life as compared to their unmarried counterpart*" is rejected. Although the interaction effects are non-significant yet the score of married jail inmates is very less compared to other groups. More appropriately, the average score of inmates on the Satisfaction with life scale was found as 13.735 whereas the average score of non-inmates was founds as 22.28. It shows that the non inmates have more satisfied as compared to their counterpart. The problem of their personal life may be important areas. Similarly the average score of married was 17.535 whereas the average score of un-married was 18.48. It shows that Unmarried have more anger as compare to their counterpart. The average score of married jail inmates on the Satisfaction with life scale was found as 12.87 whereas the average score of unmarried inmates was founds as 14.60. It shows that the unmarried inmates have more level of anger as compared to their counterpart.The unmarried jail inmates are more worried about jail life. The average score of married non-inmates on the Satisfaction with life scale was found as 22.20 whereas the average score of unmarried non-inmates was founds as 22.36. It shows that the unmarried non-inmates have slightly more level of anger as compared to their counterpart.

The average score of married inmates on the Satisfaction with life scale was found as 12.87 whereas the average score of married non-inmates was founds as 22.20. It shows that thc -married non-inmates have slightly more good level of satisfaction as compared to their counterpart. However the computed statistics has already showed the difference as statistically non-significant. The average score of married inmates on the Satisfaction with life scale found as 14.60 whereas the average score of unmarried non-inmates was founds as 22.36. It shows that the – unmarried non-inmates have slightly more level of satisfaction as compared to their counterpart. However the computed statistics has already showed the difference as statistically non-significant. Average score of married inmates was found 12.87 whereas the average score of unmarried non-inmates was found 22.36. It shows that unmarried non-inmates have more anger as compared to their counterpart. Average score of married non-inmates was found 22.20 whereas the average score of unmarried inmates was found 14.60. It shows that the married non-inmates were more satisfied as compared to their counterpart.

The main effect of Institution on the measure of PGI-health questionnaire Area (a) was found F (1, 116) = .158, p>.05 as statistically significant. There is very less difference between the two groups, therefore hypothesis No. 7 which states that *"The jail inmates will be poor in general health as compared to their non-jail inmate's counterpart"* is rejected. The main effect of Marriage was found F(1, 116) = .440, p>.05 as statistically non-significant. But the interaction between I X M was found F(1, 116) = .310, p>.05 as statistically significant. There is very less difference among the entire group. Therefore, hypothesis No. 8 which states that *"The married jail inmates will be poor in general health as compared to their counterpart unmarried counterparts"* is rejected. More appropriately, the average score of inmates on the PGI Health Questionnaire N-1 was found as 14.935 whereas the average score of non-inmates was founds as 14.68. It shows that the non inmates were healthier as compared to their counterpart. Similarly the average score of married was 14.635 whereas the average score of unmarried was 14.98. It shows that married were healthier as compare to their counterpart. The average score of married jail inmates on the PGI Health Questionnaire N-1 was found as 14.97 whereas the average score of unmarried jail inmates was founds as 14.90. It shows that the married inmates have healthier as compared to their counterpart. The average score of married non-inmates on the PGI Health Questionnaire N-1 was found as 14.30 whereas the average score of unmarried non-inmates was founds as 15.06. It shows that the married non-inmates were slightly healthy as compared to their counterpart. The average score of married inmates on the PGI Health Questionnaire N-1 was found as 14.97 whereas the average score of unmarried non-inmates was founds as 14.30. It shows that the - married non-inmates were slightly healthy as compared to their counterpart. The average score of married inmates on the PGI Health Questionnaire N-1 was found as 14.90 whereas the average score of unmarried non-inmates was founds as 15.06. It shows inmates were slightly healthier as compared to their counterpart. However the computed statistics has already showed the difference as statistically non-significant. Average score of married jail inmates was found 14.97 whereas the average score of unmarried non-inmates was found 15.06. It shows that married jail inmates have better health as compared to their counterpart. Average score of married non-inmates was found 14.30 whereas the average score of unmarried inmates was found 14.90. It shows that the unmarried inmates have more good health as compared to their counterpart.

The main effect of Institution on the measure of Wellbeing-Area-A was found $F(1, 116) = .697$, $p > .05$ as statistically significant. The main effect of Marriage was found $F(1, 116) = 1.519$, $p > .05$, as statistically non-significant. But the interaction between I × M was found $F(1, 116) = 3.492$, $p < .05$ statistically significant. There is very little difference between all the groups. Yet interaction is significant. Further the average score of inmates on the PGI Health Questionnaire N-1 was found as 7.25 whereas the average score of non-inmates was founds as 7.58. It shows that the jail inmates have better health as compared to their counterpart. Similarly the average score of married people was 7.03 whereas the average score of unmarried was 7.8. It shows that unmarred people are healthy as compare to their counterpart. Similarly the average score of married inmates on the PGI Health Questionnaire was found as 7.10 whereas the average score of unmarried inmates was founds as 7.40. It shows that the unmarried inmates have slightly better health as compared to their counterpart. Further the average score of married non-inmates on the PGI Health Questionnaire N-1 was found as 6.96 whereas the average score of unmarried non-inmates was founds as 8.20. It shows that the married non-inmates are slightly healthy as compared to their counterpart. The average score of married inmates on the PGI Health Questionnaire N-1 was found as 7.10 whereas the average score of married non-inmates was founds as 6.96. It shows that the –married jail inmates were slightly more good in healthy as compared to their counterpart. However the computed statistics has already showed the difference as statistically non-significant. The average score of married inmates on the PGI Health Questionnaire N-1 was found as 7.40 whereas the average score of unmarried non-inmates was founds as 8.20. It shows that the –unmarried non-inmates were slightly more healthy as compared to their counterpart. Similarly the average score of married inmates was found 7.10 whereas the average score of unmarried non-inmates was found 8.20. It shows that married inmates have more better health as compared to their counterpart. Further average score of married non-inmates was found 6.96 whereas the average scores of unmarried inmates was found 7.40. It shows that the unmarried inmates were healthier as compared to their counterpart.

The main effect of Institution on the measure of was found $F(1, 116) = 1.92$, $p > .05$ as statistically significant. The main effect of Marriage was found $F(1, 116) = .077$, $p > .05$ as statistically non-significant. But the interaction between I X M was found $F(1, 116) = 1.145$, $p > .05$ as statistically significant. There is very little difference between all the groups therefore hypothesis No. 7 and 8 are rejected. The average score of inmates on PGI Health Questionnaire N-1 was found as 7.66 whereas the average score of non-inmates was founds as 7.08. It shows that the inmates were healthier as compared to their counterpart. Similarly the average score of married was 7.6 whereas the average score of unmarried people was 7.15. It shows that un-married people were healthier as compare to their counterpart. The average score of married inmates on the anger self-report scale was found as 7.83 whereas the average score of unmarried inmates was founds as 7.50. It shows that the unmarried inmates were more healthy as compared to their counterpart. Further the average score of married non-inmates on PGI Health Questionnaire N-1 was found as 7.37 whereas the average score of unmarried non-inmates was founds as 6.80. It shows that the married non-inmates were slightly healthy as compared to their counterpart. The average score of married jail inmates on PGI Health Questionnaire N-1 was found as 7.83 whereas the average score of married non- inmates was founds as 7.36. It shows that the -married inmates were slightly more healthy as compared to their counterpart. Similarly the average score of married inmates on PGI Health Questionnaire N-1 was found as 7.83 whereas the average score of unmarried non-inmates was founds as 7.50. It shows that the married jail inmates were slightly healthier as compared to their counterpart. However the computed statistics has already showed the difference as statistically non-significant. Average score of married inmates was found 7.83 whereas the average score of unmarried non-inmates was found 6.80. It shows that married inmates were healthier as compared to their counterpart. Average score of married non-inmates was found 7.37 whereas the average score of unmarried inmates was found 7.50. It shows that the unmarried jail inmates were healthier as compared to their counterpart.

The main effect of Institution on the measure of spiritual wellbeing measure was found $F(1, 116) = 5.57$, $p<.01$ as statistically significant.. The jail inmates were found more religious compared to their counterpart. Therefore, hypothesis No. 9 which states that the "The jail inmates will be more religious as compared to their non-jail counterparts" is accepted. The main effect of Marriage was found $F(1, 116) = .892$, $p>.05$ as statistically non-significant. The interaction between I x M was found $F(1, 116) = .337$, $p>.05$ as statistically non-significant. Unmarried jail inmates were found more religious therefore hypothesis No. 10 which states that "*The married jail inmates will be more religious as compared to their unmarried counterpart*" is rejected.The average score of inmates on Religiosity was found as 127.58 whereas the average score of non-inmates was founds as 121.745. It shows that the inmates have more level of religious as compared to their counterpart. Similarly the average score of married was 123.315 whereas the average score of unmarried was 123.945. It shows that Unmarried have more religious as compare to their counterpart. The average score of married inmates on Religiosity scale was found as 127.13 whereas the average score of unmarried inmates was founds as 128.03. It shows that the unmarried inmates have more level of religious as compared to their counterpart. The average score of married non-inmates on Religiosity scale was found as 123.63 whereas the average score of unmarried non-inmates was founds as 119.86. It shows that the married non-inmates have slightly more religious as compared to their counterpart. However the computed statistics has already showed the difference as statistically non-significant. The average score of married inmates on Religiosity scale was found as 127.13 whereas the average score of married non-inmates was founds as 123.63. It shows that the -married inmates have slightly more religious as compare to their counterpart. However the computed statistics has already showed the difference as statistically non-significant. The average score of married inmates on Religiosity scale was found as 128.03 whereas the average score of unmarried non-inmates was founds as 119.86. It shows that the – married inmates have slightly more religious as compared to their counterpart. Average score of married inmates was found 127.13 whereas the average score of unmarried non-inmates was found 119.86. It shows that married inmates have more anger as compared to their counterpart. Average score of married non-inmates was found 123.63 whereas the average score of unmarried inmates was found 128.03. It shows that the unmarried inmates have more religious as compared to their counterpart.

The main effect of Institution on the measure of substance abuse scale was found $F(1, 116) = 2.78$, $p<.01$ as statistically significant. The non inmates reported more involvement in substances as compared to the jail inmates therefore hypothesis No. 11 which states that "The jail inmates will be more involved in alcohol, smoking, and substances as compared to their non-jail inmates" is rejected. The main effect of Marriage was found $F(1, 116) = .016, p > .05$ as statistically non-significant. The interaction between I X M was found $F(1, 116) = .898, p > .05$ as statistically non-significant. The married people reported less involvement in ASSIST compared to other group. Therefore hypothesis No. 12 which states that "*The married people will be more involved in alcohol, smoking, and substances compared to their unmarried counterpart* "is also rejected. The average score of inmates on Alcohol, Smoking and Substance Involvement Screening test was found as 4.58 whereas the average score of non-inmates was founds as 5.66. It shows that the jail inmates have more level of substance as compared to their counterpart. Similarly the average score of married was 5.43 whereas the average score of unmarried was 4.81. It shows that unmarried have more substance as compare to their counterpart. The average score of married inmates on Alcohol, Smoking and Substance Involvement was found as 4.93 whereas the average score of unmarried inmates was founds as 4.23. It shows that the unmarried inmates have more level of substance as compared to their counterpart. The average score of married non-inmates on Alcohol, Smoking and Substance Involvement was found as 5.93 whereas the average score of unmarried non-inmates was founds as 5.40. It shows that the married non-inmates have slightly more level of substance as compared to their counterpart. However the computed statistics has already showed the difference as statistically non-significant. The average score of married inmates Alcohol, Smoking and Substance Involvement was found as 4.93 whereas the average score of unmarried Himachali non-inmates was founds as 5.93. It shows that the –married non-inmates have slightly more level of substance as compared to their counterpart. The average score of married inmates on Alcohol, Smoking and Substance Involvement Screening test was found as 4.23 whereas the average score of unmarried non-inmates was founds as 5.40. It shows that the – unmarried non-inmates have slightly more level of substance involvement as compared to their counterpart. Average score of married inmates was found 4.93 whereas the average score of unmarried non-inmates was found 5.40. It shows that married inmates have more substance as compared to their counterpart. Average score of married non-inmates was found 5.93 whereas the average score of unmarried inmates was found 4.23. It shows that the unmarried inmates have more substance as compared to their counterpart.

SUGGESTIONS

1. The jail inmates be dealt with reform and humanitarian approach instead of fire-fighting approach. An opportunity be provided for basking light, safe shelter, appropriate blanket, cots, followed with sanitation and drainage facilities. Hygienic environment be provided to them. The jail inmates be provided three time healthy and nutritious food.
2. Recreational facilities be increased that include games, television, video, radio and email etc. Appropriate material be provided to them. Picnic facilities be also provided once in a year to the jail inmates.
3. The jail inmates be encouraged to pursue education, to impart professional education in order to enhance their capabilities. Library facilities be enhanced and new news bulletin, news papers, magazine, journals, books and thesis may be provided to them. Beside this religious books, and books related to psychology especially to the moral and personality development be provided to them. Workshop, seminar and conference be started so as to modify their thought process.
4. The inmates be provided job in all the sectors after completing their term. Risk intervention team and

high risk alert team may launch treatment package that in turn may prove effective for the same. Further prisoner support schemes and case management team may help a lot. Further there is a need to increase role of fellow prisoners and extension peer support.

5. New system for monitoring punishment and adjudicating grievance, prison routine, safety issues ad support services be provided to them. Prisoner handbook may be provided to them. Constructive activity within the prison be started. Support from prison staffs and probation officer be provided. Hopes and plans for future be provided.
6. Family support and visit be provided to them. Separate guest room be built near the jail where the jail inmates can stay with their family members such as spouse.
7. Inter-departmental communication be started. The staffs that are professionally trained and valued be deputed.
8. Mental health services should be integrated with Health Service Executives. Yoga, meditation, art and music such as Bhajan Kirtan therapy be encouraged through instructor. Thus diversion into non-forensic (outside) mental health services is needed to them.
9. Proper maintenance of Medical Record be kept. Regular checkup, treatment to the civil hospital even in minor diseases
10. Exchange of information with community care providers and residential care provider.Development of non-pharmacological intervention. Self-harm/suicide support and assessment program.

11. There is a dire need of Psychologists especially a counselor in order to impart counselling as well as to see the impact of treatment in order to know the changes in behaviour and thought processes. The counselor should appraise their goal, aim and objectives of the life by appropriate behavioural modification techniques.
12. The prisoner be rewarded and reinforced for their good behaviour. The term of jail should be short not lengthy because the criminal behaviour is situation specific and momentary. Once commiting the crime the criminal repent throughout the life.
13. Although some short term prisonment is necessary yet it should be converted into finance after its commitment. If the prisoner again and again (2-3 time) commits the same crime then there should be a provision of capital punishment.
14. The prison duration be short in which the inmates be dealt with humanitarian approach by forgetting fire fighting approach.
15. There should be a complaint boxes that should be tackled only by the supreme authority not by the jail staffs. In this respects action be also taken to the jail staffs.

REFERENCES

Abramsky, S. & Fellner, J. (2003). US prisons and offenders with mental illness. *http://www.org./report/2003/usa10003.pdf.*

Agali. P. O. (2004). Relative efficacy of reality therapy and assertiveness training in assisting prison inmates adjust to life after prison. *An unpublished Ph.D Thesis,* University of Ilorin, Nigeria.

Ahmad Safia., Ahmjad Hanon., & Sumboo S.S.(1988) Personality study of individuals regularly practicing transcendental meditation technique. *Journal of Personality & Clinical Studies, 4(1),* 89-92.

Aminabhavi (1996). Effect of yogic practice on attitudes toward yoga and menta health of adults. *Praachi Journal of Psycho-Cultural Dimensions, 12(2),* 117-120.

Anandamurti, S. S. (2011). Dimensions of psychic longing. *The Times of India,* 12th December.

Baumann, M., Meyer, R, Le Bihan, E, & Houssemand, C. (2008). Mental health (GHQ12; CES-D) and attitudes towards the value of work among inmates of a semi-open prison and the long-term unemployed in Luxembourg. *BMC Public Health, 8,* 214.

Baumeister, R. F., Smart, L., & Boden, J. M. (1996). *Relation of threatened egotism to violence and aggression: The dark side of high self-esteem.*

Baumeister, R. F.; Catanese, K. R., Wallace, H. M. (2002). Conquest by force: A narcissistic reactance theory of rape and sexual coercion. *Review of General Psychology, 6(1),* Mar, 92-135.

Belenko, S & Peugh, J. (2004).Estimating drug treatment needs among state prison inmates. *Handbook of labour economics, 3*(3), 3529-3571.

Belenko, S. & Peugh, J. (2004). Estimating drug treatment needs among state prison inmates. *Drug and Alcohol Dependence, 77(3-7),* 269-281.

Belenko, S. & Peugh, J. (2004). Estimating drug treatment needs among state prison inmates. *Handbook of Labor Economics, 3(3),* 3529-3571.

Bhatt, J. (2011). Inequalities prevails in Himachal too. *The Times of India.* Chandigarh: HT Press.

Bird, C. (1917). From Home to the Charge: A Psychological Study of the Soldier. The American *Journal of Psychology, 28(3),* 315-348.

Blumenthal, H. (1999). Life line: Lifers helping lifers swim not sink. *Let's Talk, 24(2),* 5-7.

Bonner, R. L. (2006). Stressful segregation housing and psychosocial vulnerability in prison suicide ideators. *Suicide Life Threat Behaviour, 36(2),* 250-254.

Bonta, J., & Gendreau, P. (1990). Reexamining the cruel and unusual punishment of prison life. In Flanagan, T. (Ed.), *Long-term imprisonment: Policy, science and correctional practice (pp. 75-94).* London: Sage Publications.

Brent A. P. & David, M. P.(1999). Structural and social psychological determinants of prisonization. *Journal of Criminal Justice, 27(5*), 427-441

Caddle, D. & Crisp, D. (1997). Imprisoned Women and Mothers, Research Study 162, Home Office Research and Statistics Directorate. University of Massachusetts Medical School, Worchester

Chandiramani K., Jena R. and Verma S. K. (1995-a). Human figure drawings of prisoners and

Chandiramani K., Verma S. K. and Dhar P. L. (1998). Psychological Effects of Vipassana on

Chaudhary, L. (1999). Effectiveness of vipassana meditation as a technique of stress management and reformation among adolescent prisoners. Unpublished dissertation.

CNN Interactive. (1998). Trend toward solitary confinement worries experts. Available: *http://cnn.com/US/9801/09/solitary.confinement/*

Coleman, J. C. (1969). Abnormal psychology and modern life. Bombay: D. B. Taraporevala & Son.

Correctional Service of Canada. (1998). *Performance report for the period ending March 31.* Ottawa: Supply and Services Canada.

Crawford, N. (2003). Helping inmates cope with prison life. *Monitor on Psychology, July-August.* The psychologist can play very important role in helping jail inmates to cope with prison life.

Dahleta, S. (2011). Jail inmates to showcase theatrical talent. *The Times of India, August 2,* pp. 4. Chandigarh: HT Press.

Day, A., Bubner, S., Jauncey, S., Williamson, P., Parker, A & Heseltine, K. (2002). Anger management and violence prevention, improving effectiveness. *Trends and Issues in Criminal Justice.* Australia: Institute of Criminal Justice.

Deepak, K. K.; Manchanda, S. K.; & Maheshwari, M. C. (1994). Meditation improves clinico- electroencephalographic measures in drug resistant epileptics. *Biofeedback & Self Regulation, 19(1),* 25-40.

Delly, C. A. (2006). *Self-harm in criminalized women.* Canada: Canadian Centre for substance abuse.

Deswal, D. (2011). Khaps reject SC observations. *The Times of India.* Chandigarh: HT Press.

Divya Himachal Bureau (2011). Qaidi Hatya mamley mein fanse chhey karamchari: Dharamshala mein chaar jail karmiyon vo do police jawano ka record jabt. *Divya Himachal 25* August.

Divya, A. (2011). Court versus Khap. *The Times of India,* May, 1, Sunday. Chandigarh: HT Press.

Divya, A. (2011). Court versus khap: The law of the land is up against a medieval midset that accepts no change-even if it is ordered by the Supreme Court. *The Times of India, Tuesday, May, 1* Sunday. Chandigarh: HT Press.

Donaldson, S. (1990). Prisons, jails, and reformatories. Available: *http://www.spr.org/*

Doris, J. J. & Lauren, E. G. (2006). *Mental health problems of Prison and jail inmates.* Washington: U. S. Department of Justice.

DSM (2000). *American Psychiatric Association Diagnostic and Statistical Manual of mental Disorder* 4th edition 298-301.

Eber, H. W. (1975). *Some psychometric correlates of inmates behaviour.* Georgia Department of Corrections, Atlanta.

Edward, P. (1985). Revenge is the mother of invention . New York: Basil Blackwell Inc., 1985. Boone, J. (1975). *U.S. Federal Prison History Unfolded,* Fortune News, May.

Elaine, M., & Gallagher, R. N. (1990). Emotional, social and physical health characteristics of older men in prison. *International Journal of Aging and Human Development, 31 (4),* 251-265.

Esere, M. O. (2011). Effectiveness of six thinking hats training in improving the psychological well being of prison inmates in Ilorin central prison. University of Ilorin, Nigeria

Fellner, J. (2006). A corrections quandary: Mental illness and prison rules. *Harvard civil rights –Civil Liberties Review, 41(1),* 22.

Feraro, K., Johnson, J. M., Jorgensen, S. R., & Bolton, F. G. (2005). Problems of Prisoners' Families: The Hidden Costs of Imprisonment. *American Journal of Public Health, 95(10),* 1725-1736.

Flanagan, T. J. (1980). The pains of long-term imprisonment. *British Journal of Criminology, 20,* 148-156.

Fong, R. S., & Buentello, S. (1990). The detection of prison gang development: An empirical assessment. Paper presented to the annual meeting of the Academy *of Criminal Justice Sciences,* in Denver, Colorado.

Freeman, R. B, (2007). The economics of crime. *Drug & Alcohol Dependence, 7,* 269-281

Freudenberg, N. (2002). Adverse effects of US jail and prison policies on the health and well-being of women of color. *American Journal of Public Health, 92(12),* 1895–1899

Freudenberg, N., Daniels, J., Crum, M., Perkins, T., Richie, B. E. (2005). *Coming home from Jail: The social and health consequences of community reentry for women, male adolescents, and their families and communities.*

Gabel, S. (1994). Behavioral problems in sons of incarcerated or otherwise absent fathers: The issue of separation. *Family Process, 31, 3,* 303–314.

Genders, E., & Player, E. (1990). Women lifers. In Flanagan, T. (Ed.), *Long-term imprisonment: Policy, science and correctional practice (pp. 117-127).* London: Sage Publications.

Green, N. (1997). Okimah Ohci healing lodge. *Let's Talk, 22(4),* 11-12.

Guy, E., Platt, J., & Bullock, S. (2007). Mental health status of prisoners in an urban jail. *Psychiatric Services, 58,* 114-120.

Hafeez, M. (2011). Arthur Rd jail inmates set the stage on fire. *The Times of India.* Chandigarh: HT Press.

Haley, H. J. (1984). Does the law need to know the effects of imprisonment? *Canadian Journal of Criminology, 26,* 479-491.

Hammersle, R., & Cregan (1986). Drug addiction and Vipassana Meditation. 'A Reader': International *Seminar,* Vipassana Research Institute, 211-214.

Haney, C. (2001). The Psychological impact of incarceration: Implications for post-prison adjustment. Presented at U.S. Health and Human Services, The Urban Institute, "From Prison to Home" Conference, January 30-31, 2002. Retrieved *from: http://aspe.hhs.gov/hsp/prison2home02/haney.pdf*

Haney, C. (2001). The psychological impact of incarceration: Implication for post prison adjustment. University of California.

Haney, C., & Zimbardo, P (1998). The past and future of U.S. prison policy: Twenty-five years after the Stanford Prison Experiment. *American Psychologist, 53(7),* 709-727.

Hartta, Y., Zinta, R. L., & Tiwari, A. K. (2009). Appraisal of HIV-AIDS consciousness among college youths in Himachal Pradesh. *Social Change, 39 (3),* 406-428.

Heney, J. (1990). Report on self-injurious behaviour in the Kingston Penitentiary for Women. *Report submitted to Correctional Services Canada.* Ottawa: Supply and Services Canada.

Herbert, I. (2007). British soldiers 'go Awol because Army ignores mental health problems. *independent.co.uk.*

Holtz, T. A. (2002). Reaching out from behind bars: The constitutionality of laws barring prisoners from the internet. Brooklyn Law School. *Brooklyn Law Review* (Spring, 2002).

Hopper, C. B. (1990). Patterns of sexual adjustment among prison inmates. *Paper presented to the annual meeting of the Academy of Criminal Justice Sciences,* March 16, in Denver, Colorado.

HT Correspondent (2011). Why are me violent? *The Times of India,* January, 30, Sunday. Chandigarh: HT Press.

Ivanoff, A.; Smyth, N. J., Grochowski, S. Jang, S., J., & Klein, K. E. (1992). Problem solving and suicidality among prison inmates: another look at state versus trait. *Journal of Consulting & Clinical Psychology, 60(6),* 970-973.

James. D. J., & Glaze, L.(2006). Mental health problems of prison and jail inmates. Bureau of Justice Statistics, Special Report.

Jin . (1992). Efficacy of Tai Chi, brisk walking, meditation, and reading in reducing mental and emotional stress. *Journal of Psychosomatic Research, 36(4),* 361-370.

John Howard Society of Winnipeg. (1990). Prisoners: An historically disadvantaged group. Winnipeg: author.

Jon F. S. & Gisli, H. G. (1996). The psychological characteristics of 'false confessors'. A study among icelandic prison inmates and juvenile offenders. *Personality & Individual Differences, 20(3),* 321-329.

Jürgens, R. (1996). HIV/AIDS in prisons: Final report. *Montreal Canadian HIV-AIDS Legal Network and the Canadian AIDS Society.*

Kamal, N. (2010). Rape of dalit: DBA on indefinite strike. *The Times of India, Tuesday,* December16, 2011. Chandigarh: HT Press.

Khurana, A. (1996). Effect of Vipassana Meditation on Quality of Life of Undertrials. In Vimla Veeraraghavan (Ed.) , *Certain Perspectives of Quality of Life.* Delhi: Krishna publishers.

Khurana, A. (1999). Vipassana meditation and subjective well-being of under trials. *Indian Journal of Criminology, 20-25.*

Klein , S. R. & Bahr, S. (1992). An Evaluation of a Family-Centered Cognitive Skills Program for Prison Inmates. Family Process, 31 (3), 303–314.

Klein, S. R. & Bahr, S. J. (2004). An evaluation of family –centered cognitive skill program for prison inmates. Canada: University of Victoria.

Krishna, Usha. (1993). Adolescent's delinquent behaviour and personality. *Indian Journal of Criminology, 21(3),* 90-94.

Kumar, G. (2011). *Apne watan nahin jana chahate Tihad mein bandh 67 Pak kaidi.* Amar Ujjala, 14th Sunday, August. Chandigarh: Press

Lauterpacht, E. & Greenwood, C. J. (2002). *International Law Reports,* Cambridge University Press.

Lea, H. C. (1991). *Witchcraft,* pg 236 as quoted in Camille Naish, *Death Comes to the maiden: Sex and execution 1431-1933* (London: Routledge, 1991), 28.

Lebel, P. (2008). An examination of impact of formerly incarcerated persons. *Journal of Offender Rehabilitation, 46, (1(2),* 1-33.

Levinson, Sanford (2006). *Torture: A Collection.* Oxford University Press, USA.

Lindquist, C. H. (2000). Social integration and mental well-being among jail inmates. *Sociological Forum, 15, 3,* 431-455

Logan, T. K., Cole, J. & Leukefeld, C. (2002). Women, sex, and HIV: Social and contextual factors, meta-analysis of published interventions, and implications for practice and research. *Psychological Bulletin, 128(6),* Nov 2002, 851-885

Lorber, M. F. (2004). Psychophysiology of aggression, psychopathy, and conduct problems: A meta-analysis. *Psychological Bulletin, 130(4),* 531-552.

Louann, B. (2006). *The Female Brain.* New York: Broadway Books.

Loveleen & Zinta, R. L. (2006). Mental Retardation: A case study. *Journal of Community Guidance & Research, 23(* 3), 296-305.

Lovell, D. & Jemelka, R. (1996). When inmates misbehave: The cost of discipline. *Prison Journal,* 165-167.

Maeve, M. & Katherine R. N. (1999). The Social Construction of Love and Sexuality in a Women's Prison. *Advances in Nursing Science, 21 (3),* 46-65.

Mahapatra, D. (2011). Socially beneficial tool turning into legal terrorism? Misuse of Section 498A has resulted in innocents get ting punished. *The Times of India, Tuesday, July 15.* Chandigarh: HT Press.

Malkin, I. (1995). The role of the law of negligence in preventing prisoners' exposure to HIV while in custody [On-line]. Available: *http://www.aidslaw.ca/elements/B5MalkE.html*

Manjul, M. & Zinta, R. L. (2010). Depression as a devil in the human development: A case study. *Journal of Community Guidance & Research, 27 (1),* 103-112.

Manjul, M., & Zinta, R. L. (2008). Motivational approaches as a predictor of strategies for utilizing forest resources to economic development. *Nava Arthiki, 16, 2A,* 1-23.

Marlene Samuelson, M. Carmody, J. Kabat-Zinn,J. & Bratt, M. A. (2007). Mindfulness-based stress reduction in Massachusetts correctional facilities. *The Prison Journal, 897(2),* 254-268.

McArthur, M., Camilleri, P. & Webb,H. (1999). *Strategies for managing suicide and self-harm in prison.* Australia: Australian Institute of Criminology.

McArthur, M., Camilleri, P., & Webb, H. (1999). Strategies for managing suicide and self-harm in prisons. *Trends and issues in criminal justice,* Australian Institute of Criminology.

Mehta, V. S. (2011). Man versus nature. *The Times of India,* February, 14, Monday. Chandigarh: HT Press.

Meyers, R., Le Behan, E. & Housemand, C. (2008).Mental health (GHQ12; CES-D) and attitudes towards the value of work among inmates of a semi-open prison and the long-term unemployed in Luxembourg. *BMC Public Health, 8,* 214.

Michelle A. L, & Steven, B. (2000). Predicting retention in a residential drug treatment alternative to prison program. *Journal of Substance Abuse Treatment, 19(2),* 145-160

Milbrath, L. (1979). Policy relevant quality of life research, *Annals, AAPSS,* 444, 32-45.

Moudgil, A. C., Verma, S.K.& Kaur, Kuldip (1986) P.G.I. Quality of Life Scale (Revised Form). *Indian J. Clinical Psychology. 13,* 175-184

Mitchell, E. W. (1999). Does psychiatric disorder affect the likelihood of violent offending? A review and critique of the major findings. Medicine, Science & Law, 39 (1), pp. 23-30.

Morgan, R. D., Steffan, J. Shaw, L. B. & & Wilson, S. (2007). Needs for and barriers to correctional mental health services: inmate perceptions. *International Journal of Prisoner Health, 3(2),*

Naish, C. (1991). *Death Comes To The Maiden: Sex and Execution 1431-1933* (London: Routledge, 1991), 27.

Napoleon, B. (1961). *Letters and Documents of Napoleon, Volume I: The Rise to Power,* selected and translated by John Eldred Howard (London: The Cresset Press, 1961), 274.

Nathan, C. Stephen, N. (1973). On being useful: The nature and consequences of psychological research on social problems. *American Psychologist, 28(3),* 199-211.

Negi, S. K. & Zinta, R. L. (2010). Psychological wellbeing among polyandrous and monogamous socially disadvantaged people of Kinnaur. *Indian Journal of Psychological Science, 1,* 21-44.

Novotney, A. (2009). Strong in mind and body. *Science Watch, 40(11),* 40-41.

Odu, B. K. (2000). The effectiveness of decision making skills in improving the psychological well-being of retirees. *Nigerian Journal of Applied Psychology, 6*(1) 80-86.

Ogloff, J. R. P., Roech, R., & Hart, S. D. (1994). Mental health services in jails and prisons: Legal, clinical and policy issues. *Law & Psychological Rev., 109.*

Onagoruwa, K. (2001). *Interview on the launching of the civil liberties organization.* Latest publication of *Behind the Wall*

Osofsky . (1996). Psychiatry behind the walls: mental health services in jails and prisons. *Bull Menninger Clin, 60(4),* 464-479.

Otto, K & Dalbert, C. (2005). *Belief in a just world and its functions for young prisoners Advances in Nursing Science.* Department of Educational Psychology, German (available on online).

Otto, K., & Dalbert, C., (2005). Belief in a just world and its functions for young prisoners. *Journal of Research in Personality, 39(6),* 559-573

Pamela, A. C.& Frank, P. D. (2000). Impact of attitudes and suicidal ideation on adolescents' intentions to seek professional psychological help. *Journal of Adolescence, 23, 1,* 35-45.

Parry, John T. (2010). *Understanding Torture: Law, violence, and political identity.* Ann Arbor, MI: University of Michigan Press.

Pelissier, B. M. M. Camp, S. D., Gaes, G. G. Saylor, W. G. & Rhodes, W. (2003). Gender differences in outcomes from prison-based residential treatment. *Journal of Substance Abuse Treatment, 24, 2,* 149-160.

Plugge, E., Douglas, N., Fitzpatrick, R. (2006). *The health of women in prison: study findings.* Department of Public Health: University of Oxford.

Pogrebin, M. R. & Dodge, M. (2001). Women's account of their prison experience: A retrospective view of their subjective realities. *Journal of Criminal Justice, 29, 6,* 531-541

Pradhan, K. (2011). Prisoner of rhetoric. *The Times of India, 24 July, p. 13.* Chandigarh: HT Press.

Reddy, Peter (2010). *Torture: What you need to know.* Ginninderra Press, Canberra, Australia.

Reynolds, N. S., Walky, H. F., & Green, D. E. (1994). The anger self-report: A psycholometrically sound (30 items) version. *New Zealand Journal of Psychology, 23,* 64-70.

Rhodes, L. A. (2005). Pathological Effects of the Super maximum Prison. *American Journal of Public Health, 95 (10),* 1692–1695.

Ryan, R. M., & Edward L. Deci, E L. (2001). On happiness and human potentials : A review of research on hedonic and eudaimonic well-being. *Annual Review of Psychology, 52,* 141-166

Ryley, S. C. (2003). *History of Torture throughout the ages.* Kessinger Publishing.

Sabbath M. J., & Cowles, E. L. (1992). Problems associated with long term incarceration. *Forum on Corrections Research, 4(2),* 9-11.

Santos, M. (1995). Facing long-term imprisonment. In Flanagan, T. (Ed.), *Long-term imprisonment: Policy, science and correctional practice (pp. 36-40).* London: Sage Publications.

Sarwar, B. (2011). Sorry, I shot your father down. *The Times of India,* Friday, August 12. Chandigarh: HT Press.

Scarry, Elaine (1985). *The body in pain the making and unmaking of the world.* Oxford: Oxford University Press.

Schmid, Alex P.; Crelinsten, Ronald D. (1994). *The politics of pain: torturers and their masters.* Boulder, Colo: Westview Press.

Selvamurthy, W. (1993). Yoga and stress management: Physiological perspectives. *Proceedings of the Indian Science Congress,* Part IV, 169.

Sen, A. (2011). Good inmates to get open air jail. *The Times of India,* 28 June, pp. 6. Chandigarh: The Times of India Press.

Sethi, N. (2011). Sonia sends food security bill to govt. *The Times of India,* Tuesday, July 8 pp. 9. Chandigarh: HT Press.

Shah, (1976). Impact of Vipassana Meditation on prisoners. College of Social Work, Mumbai (A Report).

Sharma, A., & Zinta, R. L. (2008). Alcoholism: A case study. *Indian Journal of Psychology & Mental Health, 1(2),* 195-214.

Sharma, D. K. (2011). Another Ferozepur jail inmate dies. *The Times of India, 28th* December. Chandigarh: HT Press.

Sharma,D. K. (201). Another Ferozepur jail inmate dies. *The Times of India, 28* December.

Singh, A. (2011). Education programme in Karnail Jail a hit. *The Times of India 12th* August, 2011.

Singh, A. (2011). Education programme in Karnal Jail a hit. *The Times of India, 28th*

Singh, I. P. (2011). Fraudsters prey on death row inmates families. *The Times of India, 16th* January . Chandigarh: HT Press.

Singh, I. P. (2011). Fraudsters prey on death row inmates families. *The times of India, 16th* January, 2011. Chandigarh: HT Press.

Singh, S. H. & Zinta, R. L. (2008). DHAT Syndrome: A case study. *Journal of Community Guidance & Research, 25* (2), 173-184.

Sinha, K. (2011). Compensation for clinical t rials: Just Rs. 50 Lakh for kin. *The Times of India, Friday,* August 12. Chandigarh: HT Press.

Sinha, K. (2011). Mental health care in dire straits. *The Times of India, Friday, 23* December.

Sinha, P. & Mukherjee, S. (2011). I don't control. I grab the opportunities. *The Times of India,* Tuesday, June, 6, 2011. Chandigarh: HT Press.

Smith, A., Parish, J. (2010). *When a person with mental illness goes to prison.* New York: Urban justice Centre.

Smith, P. S. (2006). The Effects of solitary confinement on prison inmates: A brief history and review of the literature. *Crime & Just. 441*

Solicitor General of Canada. (1991). *Solicitor General of Canada annual report 1990-91.* Ottawa: Supply and Services Canada.

Srivastava, G. (2011). Educating the underprivileged. *The Times of India, Friday,* Wednesday, June, 15. Chandigarh: HT Press.

Srivastava,R. (2011). India's oldest prisoner walks free at 108. *The Times of India, 19th* June. Chandigarh: HT Press.

Stableforth, N. (1997). Women Offenders. *Let's Talk, 22(2),* 4.

Steadman, H. J. & Veysey, B. M. (1997). *Providing service for jail inmates with mental disorder.* US. National Institute of Justice.

Steadman, H. J. Osher, F. C., Robbins, P. C., Brian, B. A. & Samuels, S. (2009). Prevalence of serious mental illness among jail inmates. American Psychiatric Association. : *hsteadman@prainc.com.*

Subramaniam, T. S. (2009). Of men and minds. *Frontline, 26,*

Suter, J. & Byrne, M. (2000). *Australia: Institute of Criminology.* Female offender are different from male offender: Anger as an example.

Sykes, G. (1966). *The society of captives.* New York: Atheneum.

Takkar, J. (2011). In holy city, liquor mafia operatig from juice shops. *The Times of India,* 1st May. Chandigarh: HT Press.

Takkar, J. (2011). In holy city, liquor mafia operating from juice shops. *The Times of India,* Sunday, May, 1. Chandigarh: HT Press.

Task Force on Federally Sentenced Women (1990). *Creating choices: The report of the task force on federally sentenced women.* Ottawa: Supply and Services Canada.

Task Force on Long Term Offenders. (1998). *Implementing the life-line concept: Task force on long term offenders.* Ottawa: Supply and Services Canada.

Tess, M. S. & Carl, B. C. (2010). Prison rape and psychological sequelae: a call for research. *Psychology, public policy and law, 284.* American Psychological Association.

Thakur, P. & Zinta, R. L. (2010). Appraisal of psychological frailness among rural and urban elderly people. *Research Journal Social Sciences, 18 (3),* 58-76.-

Thomas, (2011). Jail surroundings polluted: Accused. *The Times of India, 23th* April. Chandigarh: HT Press.

Thomas, S. (2011). 16 year after death in custody, kin get Rs. 4L in damages: Victim's wife died during case pendency. *The Times of India, Friday,* August 12. Chandigarh: HT Press.

Thomas, S. (2011). Jail surroundings polluted: Accused. *The Times of India, 23* April. Chandigarh: HT Press.

Thomas, S. (2011). Jail surroundings polluted: Accused. *The Times of India, 23* April, 2011. HT Press.

Times News Network (2011). It was rarest of rare crime: SC. *The Times of India,* 12 August. Chandigarh: HT Press.

Times News Network (2011). NHRC proposes. Bihar police HQ disposes; DIG report on rights abuse in Bettiah Jail trashed. *The Times of India, 28* June, page 5. Chandigarh: The Times of India Press.

Times News Network (2011). Why are men violent? Times of India, Sunday, 30 January. Chandigarh: HT Press.

Tiwari, A.K. & Zinta, R. L. (2007). Hydropower generation and economic development: A study of NJPC project in Himachal Pradesh. *Research Journal Social Sciences, 15,* 2, 136-148.

Tosh, J. (1982). *The pains of imprisonment.* California: Sage Publications.

Travis, J. & Visher, C. (1995).The effectiveness of treatment for drug abusers under criminal justice supervision. *Conference on Criminal Justice Research and Evaluation.* Washington: National Institute of Justice.

Trevor-Roper, H. R. (1969). *The European Witch-Craze of The Sixteenth and Seventeenth Centuries and Other Essays.* New York: Harper and Row.

U.S. Department of Justice (1995). The effectiveness of treatment for drug abusers under criminal justice supervision. Research Report, Conference on Criminal Justice Research and evaluation, U.S.

Unnithan T. K. N., & Ahuja, R. (1977). Prisoners' meditation camp: A sociological analysis. The *Maha Bodhi Journal, 85,* 303-309.

Van Marle, H. J.C. (2007). Mental health care in prison: How to manage our care. *Journal of Research in Personality, 39, 6,* 559-573

vanMarle, H. J. C. (2007). Mental health care in prison: How to manage our care. *Psychiatric Service, 58,* 1181-1186.

Waldron, Jeremy; Colin Dayan (2007). *The Story of cruel and unusual..* Cambridge, Mass: MIT Press.

Weekes, J. R. (1992). Long-term offenders in Canada. *Forum for Corrections Research, 4(2),* 3-7.

WHO, 1999). *Mental health promotion in prison.* Http://www.euro.who-int.document/E64328pdf.

Wormith, J. S. (1984). *What do we know about long term incarceration?* Report prepared by senior research officer, Research Division, Ministry of the Solicitor General of Canada. Ottawa: Supply and Services Canada.

Wormith, J. S. (1985). Long term incarceration: Data and reason meet ideology and rhetoric. *In Solicitor General of Canada User Report: Long term incarceration: The continuing debate.* Ottawa: Supply and Services Canada.

Yiengprugsawan , V., Seubsman , S. A., & Sleigh , A. (2010). Health, well-Being, and social indicators among monks, prisoners, and other adult members of an Open University cohort in Thailand. *Journal of Religion & Health, 85, 6-27.*

Youman, K., Drapalski, A., Stuewig, J., Bagley, K. & Tangney, J. Race (2010). Differences in psychopathology and disparities in treatment seeking: community and jail based treatment seeking pattern. *Psychological Services, 7(1),* 11-26.

Young, D. (1989). Impacts of perceived legal pressure on retention in drug treatment. *University of Maryland, College Park.*

Zamble, E., & Porporino, F. J. (1988). *Coping, behaviour and adaptation in prison inmates.* New York: Springer-Verlag.

Zhang,J., Grabiner, V. E., Virginia, E., Y., Zhou & Li, N. (2010). Suicidal ideation and its correlates in prisoners: A comparative study in China. *The Journal of Crisis Intervention and Suicide Prevention, 31(6),* 2010, 335-342

Zingraff, M. T. (1975). Prisonization as an inhibitor of effective re-socialization. *Criminology, 13,* 366-381.

Zinta, R. L. & Negi, S. K. (2011). To compare the polyandrous and monogamous socially disadvantaged people of Kinnaur district on various parameter of psychological wellbeing. *Research Journal Himachal Pradesh University, 1(1),* 84-101.

Zinta, R. L. & Tiwari, A. K. (2006). Relationship between self-efficacy, socio-economic status and performance in secondary students. *Research Journal Social Sciences, 14, (1),* 121-137.

Zinta, R. L. (2006). Effect of self-efficacy, test-anxiety and short-term intervention on problem solving: A preliminary analysis. *Social Science International, 22, 1,* 3-8.

Zinta, R. L. (2006). Is guided mastery helpful in problem solving. *Psychological Studies, 51(4),* 323-325.

Zinta, R. L. (2007). Importance of eclectic intervention for the management of anxiety. A *Paper Presented to the Orientation Programme-81* in UGC Academic staff college in the vide publication No. 283/2007.

Zinta, R. L. (2008). Andheri surang mein roshni failani ki pratibadta. A book Review of Varun Kumar Tiwari. *Kritika, 1,*11,152.

Zinta, R. L. (2008). Effectiveness of guided mastery treatment for reducing test-anxiety among self-efficacious students. *Journal of Indian Academy of Applied Psychology, 34,* 2, 233-239.

Zinta, R. L. (2008). Socio-economic status as a panacea for mental health: A note on displaced families of Bhakra Dam. *Nava Arthiki, 16, 2A, 49-74.*

Zinta, R. L. (2009). Bharat mein uch shiksha ki dasha aur disha. *Sanchar Bulletin, 1(1),* 104-106.

Zinta, R. L. (2009). Different facets of quality of life among the populace of mounting India in 21st century. *Akhil Bhartiya Shodh Seminar on Iksavin sadi mein Bharat: Dasha avam disha,* organized by University Grant Commission, New Delhi. Dayanand Vedic Snatkotar Mahavidyalaya, Urai, Jalon, U. P.

Zinta, R. L. (2009). Environment and its impact on human mental health. *A Research Paper, presented in RC, 2009 on Environmental Awareness in the UGC Academic Staff College Shimla* vide publication No. 331/2009.

Zinta, R. L. (2009). Rejuvenating inner resources of the youths through psychological interventions. *A paper presented to Youth Development Counselling Centre, H. P. University, Summer Hill, Shimla*-171005.

Zinta, R. L. (2010). Dimensions of quality of life in the tribal areas of Kinnaur: A caste-wise analysis. *A Research Project proposal approved by Institute of Tribal Research and Studies, H. P. University, Summer Hill,* Shimla-171005.

Zinta, R. L. (2010). Performance among high and low self-efficacious students. In P, Ramalingam (Ed.), *Recent studies in school psychology.* Delhi: Authorspress.

Zinta, R. L. (2010). *Psychology Manual.* New Delhi: H. G. Publications.

Zinta, R. L. , Manjul, M. , Kumar, D. & Thakur, R. (2010). Assessment of the socio-economic status, depression and coping styles among single women of tribal areas of Himachal Pradesh. *Nava Arthiki, 18(2),* 72-88.

Zinta, R. L., & Tiwari, A. K. (2007). Impact of guided mastery treatment on self-efficacy among test-anxious students. *Journal of Psychological Researches, 51*(2), 91-101.

Zinta, R. L., & Tiwari, A. K. (2008). Psychological analysis on project affected and likely to be affected families of SJVN Ltd. of Himachal Pradesh. *A project Reported submitted to the Integrated Institute of Himalayan Studies, HPU Shimla*-171005.

Zinta, R. L., & Tiwari, A. K. (2008). Socio-economic and infrastructure facilities for sustainable rural development: A factorial analysis on Himachal Pradesh. A project Reported submitted to the *Integrated Institute of Himalayan Studies,* HPU Shimla-171005.

Zinta, R. L., Hartta, Y. S. & Nadda, K. (2010). Study of the attitude and opinion of college students towards HIV positive people. *Nava Arthiki, 18(2),* 89-105.

Zinta, R. L., Manjul, M., & Kumar, D. (2010). Social status and emotional wellbeing among employed and unemployed tribal single women. *Nava Arthiki, 18(1A),* 1-27.

Zinta, R. L., Negi, S. R. & Thakur, R. (2010). Caste and efficacy as a precursor of cognitive abilities: A study on tribal high school students of District Kinnaur. Research Journal *Social Science, 18(1),* 130-157.

APPENDICES

A variety of torture instruments including, at right, the iron maiden of Nuremberg.

Middle Ages

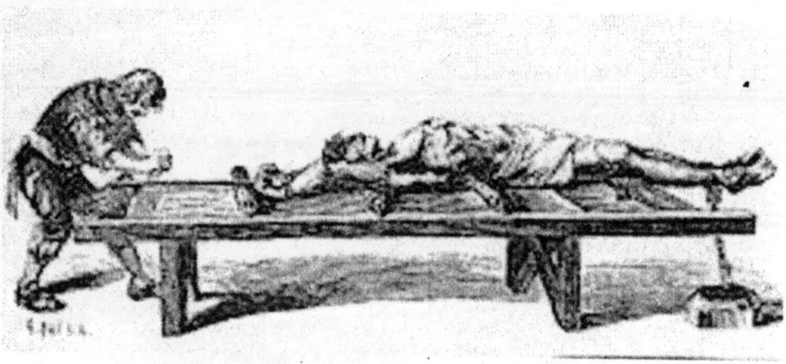

Medieval torture rack

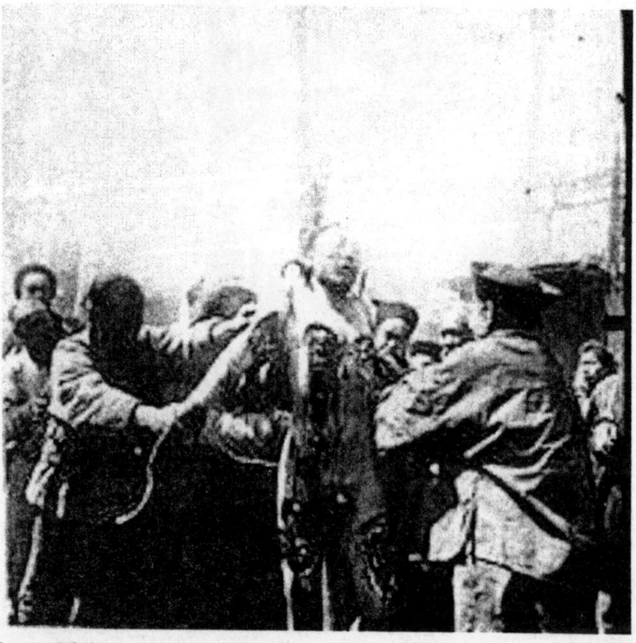

Ling Chi – execution by slow slicing – in Beijing around 1904.

Native Americans scalping and torturing prisoners, an 1873

Torture in the 16th century

Gloucester Country Asylum established in 24.7.1823

Oxfordshire and Berkshire County Asylum opened in 1.8.1846

Cane Hill Asylum opened in 1883 that was enlarged for the 2000 population.

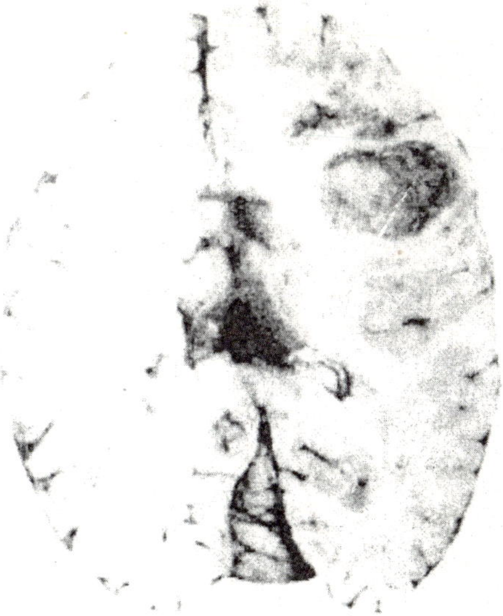

1890s: Cerebral lesion and trephenology practice in at Colney Hatch Asylum

Picture postcard of County Asylum Annexe Lancaster postmarked Lancaster 6.15 pm 25.10.1904. One of several asylum postcards in the collection of Nigel Roberts

Picture postcard of Napsbury (opened 1905) that is thought to date from the first world war, when Napsbury was a war hospital

Napsbury was a war hospital

The picture (from the collection of the Henry Ey Foundation) shows the opening ceremony the First World Congress of Psychiatry in the Grand Amphitheatre of the Sorbonne, Paris (external source) 1953

On Sunday 4.5.2003 Gordon Tozer took a break from his work on the world of asylums to photograph what is left of Coney Hill Mental Hospital in loucestershire

death penalty news